ALSO BY ALBERT JACK

Red Herrings and White Elephants
Black Sheep and Lame Ducks
Phantom Hitchhikers
Pop Goes the Weasel

LOCH NESS MONSTERS
and
RAINING FROGS

LOCH NESS MONSTERS
and
RAINING FROGS

THE WORLD'S MOST PUZZLING MYSTERIES SOLVED

Albert Jack

ILLUSTRATIONS BY SANDRA HOWGATE

RANDOM HOUSE TRADE PAPERBACKS

NEW YORK

Published in the United States by Random House Trade
Paperbacks, an imprint of The Random House Publishing
Group, a division of Random House, Inc., New York.

RANDOM HOUSE TRADE PAPERBACKS and colophon are
trademarks of Random House, Inc.

Originally published in hardcover in the United Kingdom
as *Albert Jack's Ten-Minute Mysteries* by Penguin Reference,
a member of Penguin Books Ltd.,
London, in 2007.

ISBN 978-0-8129-8005-9

Printed in the United States of America

www.atrandom.com

2 4 6 8 9 7 5 3 1

Book design by Simon M. Sullivan

0 3 1
J

This book is dedicated to

....
(insert name here)

Contents

Introduction / xi

The Famous Aurora Spaceship Mystery / 3

Try to See It from My Angle: The Bermuda Triangle / 12

Does Bigfoot Exist? / 25

The Spine-chilling Tale of the Chase Vault / 39

The Real-life Agatha Christie Mystery / 48

*Committing the Perfect Crime:
The Mysterious D. B. Cooper / 55*

Who Really Makes Crop Circles? / 66

*John Dillinger: Whatever Happened to
America's Robin Hood? / 77*

The Missing Navy Diver / 80

The Dover Demon / 88

*The Mysterious Disappearance of
the Lighthouse Keepers of Eilean Mor / 90*

Fairies at the Bottom of the Garden / 101

The Mystery of Our Lady of Fatima / 103

What Happened to the Lost King of France? / 114

The Strange Case of Kaspar Hauser / 117

The Great Loch Ness Con Trick / 121

Will the Real Paul McCartney Please Stand Up / 128

The Magnetic Strip / 136

Whatever Happened to the Crew of the **Mary Celeste**? / 138

The Men Who Cheated Death / 153

Not in the Mood: The Real Glenn Miller Story / 158

The Mystifying Death of a Media Mogul / 168

The Real Da Vinci Code: The **Mona Lisa** *Debate* / 178

*If Gentlemen Prefer Blondes,
Who Killed Marilyn Monroe?* / 184

The Piano Man / 197

The Dreadful Demise of Edgar Allan Poe / 201

It's Raining Frogs / 214

The Terrifying Affair of Spring-heeled Jack / 217

Beware of USOs / 221

The St. Valentine's Day Massacre / 227

The World's Strangest Unsolved Crimes / 233

Acknowledgments / 237
Index / 239

Introduction

We all love a good mystery, don't we? And by all, I mean each and every one of us is, or will be, captivated at one time or another by a decent real-life mystery, either one of the world's best or something on a smaller scale, such as the baffling question of why the best-looking girl is going out with a loser (again). And so, after writing my last book, *Phantom Hitchhikers* (on urban legends), and inspired by the legend of the Beast of Bodmin Moor (described in that book), I started looking at some other famous mysteries, ones that continue to fascinate us. The story of the Beast of Bodmin Moor is an example of an urban legend that could also be properly researched as a genuine mystery, and the same could be said for various other topics covered in the book. There is clearly a crossover between an urban legend and a full-scale mystery. Mysteries are fact-based, of course, and tend to be longer and more complicated; indeed, some, such as the Bermuda Triangle, UFOs, crop circles, or the Loch Ness Monster, easily provide enough material for a whole series of books.

But writing a book on just one of these would have been relatively easy. The challenge came from researching lots and lots of them and then condensing them down in a way that I know you, the readers, who continue to pay my wages, enjoy reading. And that is short, sharp, informative sections you can read on the train, in the bus queue, or while waiting to pick up the kids after you arrive at school to find them in detention. In other words, the challenge was to explain the longer mysteries in a way you can enjoy and absorb in about ten minutes. Inevitably some information will be missing, for which I apologize in advance. But the missing information isn't critical to the basic story; the "core" details of the mystery in question should all be there. In some cases, such as the sections on the Bermuda Triangle, Bigfoot, and crop circles, there are literally thousands of examples that I could have used, of course, but in each case I've kept them down to just a handful.

Another challenge has been which mysteries to select out of the scores of famous stories that exist. I have tried to be as diverse as possible, including mysterious disappearances (such as the lighthouse keepers of Eilean Mor or the crew of the *Mary Celeste*) and deaths (Marilyn Monroe and Robert Maxwell), famous crimes (the St. Valentine's Day Massacre or D. B. Cooper jumping out of a jet with $200,000 in cash), science (UFOs and USOs—science fact or science fiction?), history (the tale of the "lost dauphin") and the arts (the *Mona Lisa* and Edgar Allan Poe), ranging from the obscure (the "Dover Demon") to the world-famous (the disappearance of Glenn Miller). For example, did you know that sometimes it rains frogs or fish, or that the *Mary Celeste* was nowhere near the Bermuda Triangle, despite several claims to the contrary?

And this brings me to an important point. Given that I am a fan of the unknown and the unexplained, I have not set out to be a mystery buster in this volume. Instead I just wanted to tell the story, provide some little-known detail, and offer a rational explanation wherever I could. I wanted to provoke a bit of thought and conversation, but leave you to decide the answer for yourselves: Does the story remain a mystery, in your view, or have you managed to piece together a theory of your own? To be fair, I must admit there are some cases where I just couldn't resist presenting my own ideas and giving full rein to my skepticism, but don't let that stop you enjoying yourselves.

The truth is that the mind can play tricks on us. We know this is the case; it is why we enjoy marveling at an optical illusion or a magician's skillful sleight of hand. There are other occasions, however, when we don't (or won't) acknowledge that we have been deceived: we believe we can see something and, even though we know that it doesn't actually exist, we can still see it—because we want to. Perhaps that is why there are still so many sightings of the Loch Ness Monster. A lump of wood floating innocently on the surface of Loch Ness is immediately classified as a "sighting," while the very same lump of wood goes totally unnoticed when tossed into a less famous loch nearby.

Some things remain mysterious, of course, such as unsolved crimes and disappearances (the fate of the crew of the *Mary Celeste* remains a mystery to this day); or ghostly goings-on (no one has come up with a satisfactory explanation for the frightening disturbances that took place in the Chase vault). Where there is insufficient evidence a mystery will surely arise, but nonetheless we like to blind ourselves to this sometimes, preferring the reason to be strange and otherworldly rather than clear and matter-of-fact. The crew of the *Mary Celeste* didn't abandon ship because it was about to sink or catch fire, but because a giant squid snatched them up in its writhing tentacles, or a passing UFO swooped them away to another planet. And why should there always be a boring, down-to-earth answer for everything? A bit of mystery makes life much less dreary and infinitely more exciting.

Then there are those things that mystify us but should really be explained, such as what makes that loser so irresistible to women or why *Big Brother*—a program dependent on an audience of boring people with nothing better to do than sitting around in a room in a house watching the same sort of people doing the same thing on television—remains so popular. Or, for that matter, how Russell Brand gets to be on television. You see what I mean; some things should really be explained.

As I wrote this book I had a number of imaginary readers sitting at my shoulder. The first was you, of course, who above all want to be entertained. The rest were representatives from the groups of people who passionately believe in a particular topic, whether Bigfoot, UFOs, or the Loch Ness Monster. While our views will inevitably differ, I have still tried to be as sensitive as possible. Well, I tried. And that is the reason I gave up on my efforts to investigate the Mystery of God and the Mystery of the Mind of the Modern Woman. With the first I was on a wild goose chase, and with the second I realized it was not a subject for a ten-minute mystery. More like a ten-year mystery.

Mystery in death, as in life, has a lot going for it, and there are a lot of mysterious deaths or disappearances in this book. It's true that I once replied to the question "What would you like written on your gravestone?" with "Here lies Albert Jack, aged 287." But would "Here lies Albert Jack, aged 28" have been better—tragically cut off in my prime—or maybe no gravestone at all because I had vanished without a trace? Wouldn't it be better to be remembered for dying in mysterious circumstances after your helicopter crashed into the side of Table Mountain, upside down, than for sitting in a pool of urine, covered in biscuit crumbs in an old people's home? At least that way your children would have a good story to pass on to future generations. That way others can wonder for years what really happened to you.

I don't want to appear a mystery buster, because I love a good mystery as much as the next person. But a healthy dose of skepticism never goes amiss when tackling any of the world's mysteries. For example, when researching the Bermuda Triangle I considered the question "Who has the most money to gain or lose in the area of the Bermuda Triangle?" No, not the storyteller, nor the TV documentary maker, nor the tour operator either. It must be the marine insurance companies who would have the most at stake if mysterious forces were at work down Bermuda way. And so the first place I turned to was Lloyd's of London. Such an approach, I have always found, is the best way to separate fact from fiction, myth from mystery.

I hope you enjoy these mysteries and some of the alternative theories that I have put forward. If I come across as overly opinionated from time to time, blame it on all the research and my getting caught up in the subject. So, if you are a passionate believer in UFOs or crop circles, please don't take what I've written too much to heart and send round Reg Presley or David Icke to set fire to my trousers.

ALBERT JACK
Cape Town
July 2007

LOCH NESS MONSTERS
and
RAINING FROGS

The Famous Aurora Spaceship Mystery

Did a UFO really crash in a small town in Texas over a century ago?

When it comes to spaceships and little green men from Mars, most people's thoughts turn to the notorious events at Roswell, New Mexico, where in 1947 the U.S. government apparently captured an alien who had crashed his flying saucer. U.S. military personnel are then said to have quickly sealed off the area, removed all evidence, and engaged in a complete cover-up.

After a thorough debriefing, presumably in sign language, the little green man sadly died. Much later, the film of the top-secret autopsy supposedly carried out on him was sold on the black market, ending up nearly fifty years later, in 1995, on a prime-time TV documentary broadcast around the world. This program, *Alien Autopsy*, caused a sensation and "Martiangate" was back on the agenda with a vengeance. As is often the case, those who wanted to believe such a story inevitably did, while those of us really living on planet Earth could smell a rat. In fact, there were rats everywhere.

But it took eleven years before the program maker, Ray Santilli, admitted that the autopsy had been staged, for the most part, in a flat in Camden Town, London. Strangely enough, he owned up to this two days before a humorous parody of his subject was due to be aired on television. He confirmed that his props had included sheep brains set in jelly and knuckle joints and chicken entrails bought from Smithfield meat market.

That should have knocked the Roswell mystery on the head for good, and all those UFO enthusiasts who had been obsessing about the whole affair for years should now be quietly licking their wounds in their garden sheds, or wherever it is they go to study their favorite subject.

But Roswell wasn't the first time: aliens had been captured before. In 1897, Aurora, a small, unremarkable town near Dallas, Texas, became the site of an astonishing event.

On April 17 that year, ten-year-old Charlie Stevens was sweeping his backyard when he looked up to see smoke trailing from a large silver airship flying overhead toward Aurora. Soon after it had flown out of sight, he heard an explosion and saw a thick plume of smoke rise into the air. He was about to rush off to see what had happened when he was stopped by his father, who told him he had to finish his chores first. Just imagine that something truly momentous has just happened right in your sleepy little town: a strange airborne vehicle—something you have never seen before, maybe even a craft from another planet—crashes just a few hundred yards away from your own back gate and you are told: "Nope. You finish sweeping that there yard first, boy, and then come inside and help your ma with the breakfast."

In fact Charlie wasn't allowed to go at all. According to him, it was his father who went into town and saw the wreckage scattered about the place. Mary Evans, aged fifteen at the time, also claimed to have witnessed the crash, but stated that her parents wouldn't allow her to visit the scene either.

As S. E. Haydon reported in *The Dallas Morning News:*

About 6 o'clock this morning the early risers of Aurora were astonished at the sudden appearance of the airship which has been sailing around the country.

It was traveling due north and much nearer the earth than before. Evidently some of the machinery was out of order, for it was making a speed of only ten or twelve miles per hour and gradually settling toward the earth. It sailed over the public square, and when it reached the north part of town it collided with the tower of Judge Proctor's windmill and went to pieces with a terrific explosion, scattering debris over several acres of ground, wrecking the windmill and water tank and destroying the judge's flower garden.

The pilot of the ship is supposed to have been the only one aboard, and while his remains were badly disfigured, enough of the original has been picked up to show that he was not an inhabitant of this world.

Curiously, this story did not make even the front page. Instead it was buried on page five along with several other reports of UFO sightings. It would appear the flying saucer crash at Aurora was not particularly shocking in 1897—run-of-the-mill, you might say (in more senses than one)—even if it did destroy Judge Proctor's flower garden.

The story then told by the people of the town is that the Martian pilot, as he was termed, was given a decent Christian burial in the town cemetery and his grave marked with a single stone. The remains of the spaceship were taken away to an unknown location by the authorities and the smaller pieces were thrown into Judge Proctor's well. No other newspaper covered the story and, amazingly, the alien's resting place in the Aurora cemetery went unremarked for nearly eighty years, the small town settling back into obscurity.

That was until 1973, when the founder of the International UFO Bureau, Hayden Héwes, announced to United Press International that a grave in a small north Texan cemetery contained the body of an 1897 "astronaut" whom the report at the time had identified as being "not . . . of this world."

Newspapers all over America took up the story, and interest in the alien grave rapidly gathered pace. Curiously, as the press hounds sniffed around Aurora, they found very few residents willing to discuss the events of 1897, but despite their reticence the town soon became a hive of activity as alien hunters from around the world descended en masse.

The International UFO Bureau claimed to have found traces of radiation at both the crash site and the grave, on top of which, they said, the grass glowed red. But they were soon barred from the graveyard by local administrators, who adamantly refused to allow them to start digging around. When the investigators attempted to obtain a court order to exhume the body, the small headstone marking the grave was removed and state troopers were placed at the gates of the cemetery to prevent unauthorized access.

Hayden Hewes, interviewed for a television documentary on the subject, condemned these actions as irresponsible, stating that there was now no way of locating the grave—a site, he claimed, that was of national importance. Interestingly, Bureau representatives have never explained why they didn't just walk around looking for the red patch they had found only weeks earlier. Abandoning the grave, they turned their attention instead to Judge Proctor's farm, now under different ownership.

In 1945, Rollie Oats (yes, his real name) had bought the place. He had removed the pieces of spaceship and cleaned out the well so that his family could drink the water. Twelve years later he developed severe arthritis in his hands and, convinced the well water was responsible, had it sealed over with a six-ton slab of concrete.

During the 1973 investigation, metal found on the farm was analyzed at a laboratory, its name never disclosed, and found to be of a unique composition that could only have been produced by a very sophisticated refining process far in advance of what was possible in the 1970s, let alone the 1890s. This was held up

as hard evidence of spaceship material, and the UFO community howled for the government to reveal any information they had. In response the government ridiculed the amateur investigation, describing the Aurora spaceship story as a hoax. But of course they would say that, eh, UFO fans?

Today, amid renewed calls for a full inquiry and a thorough search of Aurora using the latest technology, some town elders now claim that the U.S. military returned many years ago, back in the 1940s, and removed all trace of the spacecraft and its pilot. Others enigmatically refuse to talk about the incident at all. One elderly resident was interviewed for the television documentary in 1973 and clearly stated on camera that the whole affair had been true. (I saw it myself, and she said it all right—there's no doubt about that, at least.) Her parents, she insisted, went to check the wreckage of the spacecraft and then told her all about it. But later, her great-granddaughter revealed she had been told the whole thing was a hoax and was puzzled why her great-grandmother would appear on camera to claim the accident had really taken place. The lure of the dollar, possibly?

But if it was all a hoax, why play such an elaborate prank in the first place, let alone keep it up for over a century? There is one very good reason—to do with the town of Aurora itself. In the middle of the nineteenth century, Aurora had been a busy, bustling trade center with a growing population and two schools. During the early 1890s, the Burlington Northern Railroad had been planning to build a route through Aurora to join the Western Railroad, when disaster struck the town in the shape of spotted fever (a form of meningitis). As the new cemetery began to take in more and more residents, the town was sealed off and people were confined to their homes.

As a consequence, the railway abruptly stopped twenty-seven miles short of the town, construction never to be resumed, and Aurora's business was devastated. Things became even worse when its major crop, cotton, was ruined by boll weevil infestation. Its fate was finally sealed by a fire that destroyed a major part of the town. All this, within the space of a few short years, left Aurora facing ruin—that is, of course, until the spaceship conveniently flew into town. The resulting (albeit somewhat delayed) publicity led to Aurora, eighty years on, being declared a place of special interest and becoming one of the most famous towns in Texas, with legendary status among the worldwide UFO community. Even today it is rumored that any unusual pieces of metal found locally are quickly confiscated by the authorities and mysteriously lost or accidentally destroyed.

One of the things that have always struck me about UFO sightings is how they always reflect the era they are reported in. For example, today we have gray aliens with oversize heads who communicate telepathically, like the alien constructed for the Roswell hoax. During the 1970s all spacemen looked like the cast of *Star Trek,* and prior to that they dressed like Buck Rogers, complete with laser guns, and got in and out of their flying saucers by ladder.

So call me cynical, but when we hear of an interred alien whose cigar-shaped spacecraft crashed into a windmill in 1897, we don't need to look too far to find out that cigar-shaped airships were first conceived in the 1890s and by 1897 were flying all over America, to the astonishment of country folk, some of whom hadn't even seen a train before.

And Aurora was far from the only location for such sightings, as soon afterward alien encounters were reported all over the U.S. Some people even ludicrously claimed they had been paid by aliens, in dollars, for spare parts for their space machines.

So imagine the scene with me. In 1897, old Farmer Gilly is standing out in his field raking the soil when a being from outer space strolls up. "Greetings, Earthling," he intones in that robotic style favored by aliens the universe over. "The satellite navigation control system on my intergalactic hyperspace craft is up the spout. Do you have anything to repair it?" Farmer Gilly looks him up and down, takes off his hat and wipes the sweat from his forehead with a shirtsleeve. "Sure thing, buddy," he replies. "Cosmic navigation broken down, has it? Probably explains why you're in Arkansas, son. Can't think of no darned good reason why else you'd be all the way out here. Let's go and see what we've got for you in that chicken shed over there." Presumably the alien pays in dollars for a roll of rusty hog wire and is on his way back to Mars by sundown. Perhaps he even takes an old hoe with him too—as a souvenir. Now, you can believe that if you want to . . .

But why jump to the conclusion that it was a spaceship that had crashed? Even back in 1897, before planes were invented (or at least ones that could fly very far), there could have been an alternative, rather more plausible explanation. Flying over Texas, an early airship, not unlike a zeppelin—or, for younger readers, the Goodyear blimp—might have sprung a leak and lost altitude. It might then have crashed into Judge Proctor's windmill and destroyed his flower bed. The resulting explosion would have melted the metal framework, which would then have re-formed into new and unrecognizable shapes when it cooled. The poor pilot might have lost his limbs in the explosion and ended up burned to a crisp, so that he didn't look human anymore. But no one in the UFO community would have bought this rather more down-to-earth explanation. Hayden Hewes can still now be seen on several television documentaries standing wistfully outside the cemetery or pictured pointing forlornly at the well, no doubt wondering how he is going to remove the six tons of concrete slab that stands between himself and his place in history.

The final word on the Aurora spaceship crash should go to the man who had the very first word, journalist S. E. Haydon. Years later Haydon, a notorious practical joker, admitted he had simply made up the story in an attempt to draw attention to the plight of his hometown and to help the dying community. He certainly did that—even if publicity took some decades to arrive—as Aurora, the town we would otherwise never have heard of, is still talked about throughout the UFO-hunting community as one of the most famous sightings of all time. They should put up a statue of him in the town square in Aurora, if there is a town square, that is.

Most UFO encounters can be explained as optical illusions, natural phenomena, meteors, or hoaxes, but a good many remain unexplained. In cases of alien abduction, it is interesting to read reports of victims who have been hypnotized and who describe their ordeals in great detail while under hypnosis. Yet when we compare these reports with those of volunteers who do not claim alien abduction, but instead are asked simply to imagine it, their recollections under hypnosis are almost exactly the same. I think this says more for the power of the imagination than it does for the likelihood of alien encounters, but then again, ours is a big universe. Infinite, in fact. Only a fool would completely rule out the idea of life on other planets in other solar systems, the closest of which are so far away they would take us seventy-five thousand years to get to in the fastest craft we currently have, which means unless aliens visit us (and possibly they do—see "Beware of USOs," page 221), then you and I will never know if there is life out there. Maybe, just maybe, we are not alone after all . . .

Try to See It from My Angle:
The Bermuda Triangle

·······································

*What is it about this infamous stretch of ocean
(and sky) that causes ships and planes
to vanish without a trace?*

At ten past two in the afternoon of December 5, 1945, five U.S. Navy Avenger torpedo bombers took off from the naval air station at Fort Lauderdale, Florida. The commander of Flight 19, Lieutenant Charles Taylor, had been assigned a routine two-hour training flight of fifteen men on a course that would take them out to sea sixty-six miles due east of the airbase, to the Hen and Chickens Shoals.

There the squadron would carry out practice bombing runs, then fly due north for seventy miles before turning for a second time and heading back to base, 120 miles away. Their plotted flight plan formed a simple triangle, straightforward to execute, and Lieutenant Taylor and his four trainee pilots headed out into the clear blue sky over a calm Sargasso Sea. Even though everything seemed set fair, some of the crew were showing signs of anxiety. This was not unusual during a training flight over open water. Less usual was the fact that one of the fifteen crewmen had failed to show up for duty, claiming he had had a premonition that something strange would happen on that day and that he was too scared to fly.

And, within a few minutes after takeoff, something strange did happen. First, Lieutenant Taylor reported that the sea appeared white and "not looking as it should." Then, shortly afterward, his compasses began spinning out of control, as did those of the other four pilots, and at 3:45 P.M., about ninety minutes after takeoff, the normally cool and collected Taylor contacted Lieutenant Robert Cox at flight control with the worried message: "Flight Control, this is an emergency. We seem to be off course. We can't make out where we are."

Cox instructed the pilot to head due west, but Taylor reported that none of the crew knew which way west actually was. And that too was highly unusual, as even without compasses and other navigational equipment, at that time of day and with the sun only a few hours from setting, any one of them could have used the tried-and-tested method of looking out of the window and following the setting sun, which will always lie to the west of wherever you find yourself.

Just over half an hour later, Taylor radioed flight control again, this time informing them he thought they were 225 miles northeast of base. His agitated radio message ended with him saying, "It looks like we are . . . " and then the radio cut out. By then they would have been desperately low on fuel, but the five Avengers had been designed to make emergency sea landings and remain afloat for long enough to give the crew the chance to evacuate into life rafts and await rescue.

A Martin Mariner boat plane was immediately sent out to assist Flight 19 and bring the men back; but as it approached the area in which the stricken crew were thought to have been lost, it too broke contact with flight control. None of the aircraft and none of the crew were ever found, and the official navy report apparently concluded that the men had simply vanished, "as if they had flown off to planet Mars." To this day, the American military has a standing order to keep a watch for Flight 19, as if they believe it was caught up in some bizarre time warp and might return at any time.

At least, that is how the story goes. And it would have had a familiar ring for some, as it wasn't the first time a mysterious disappearance had been reported in the area. On March 9, 1918, the USS *Cyclops* left Barbados with a cargo of 10,800 tons of manganese (a hard metal essential for iron and steel production) bound for Baltimore. The following day, Lieutenant Commander G. W. Worley, a man with a habit of walking around the quarterdeck clad in nothing but his underwear and a hat and carrying a cane, reported that an attempted mutiny by a small number of the 306-man crew had been suppressed and that the offenders were below decks in irons. And that was the last anybody ever heard from Worley or any of his crew. The twenty-thousand-ton *Cyclops* simply vanished from the surface of the sea, into thin air.

The conclusion at the time was that the ship had been a victim of German U-boat activity, but when investigations in Germany after the end of the First World War revealed that no U-boats had been located in the area, that theory was ruled out. Instead, speculation ranged from the suggestion—proffered quite seriously—by a popular magazine that a giant sea monster had surfaced, wrapped its tentacles around the entire ship, dragged it to the ocean bed, and eaten it, to the rumor, with UFO hysteria in full swing (see "The Famous Aurora Spaceship Mystery," page 3), that the vessel had been lifted, via giant intergalactic magnets, into outer space.

And then, in 1963, eighteen years after the disappearance of Flight 19, it happened again. The SS *Marine Sulphur Queen* was on a voyage to Norfolk, Virginia, from Beaumont, Texas. On February 3, the ship radioed a routine report to the local coast guard to give her position: she was, at the time, sailing close to Key West in the Straits of Florida. Shortly afterward she vanished. Three days later the coast guard, searching for any sign of the missing vessel, found a single life jacket floating in the sea.

Since then, no other evidence of the *Marine Sulphur Queen,* its cargo, or the thirty-nine-man crew has ever been found.

Back in 1950, connections had already been made between the disappearance of Flight 19 and of the USS *Cyclops:* reporter E.V.W. Jones was the first to suggest mysterious happenings in the sea between the Florida coast and Bermuda. Two years later, *Fate* magazine published an article by George X. Sand in which he suggested that the mysterious events—thousands of them, by his calculation—had taken place within an area that extended down the coast from Florida to Puerto Rico and in a line from each of these to Bermuda, creating what he called a "watery triangle." His views were shared by one Frank Edwards, who published a book in 1955 called *The Flying Saucer Conspiracy* in which he claimed that aliens from outer space were also operating in the same area; hence the sky was incorporated into the "watery triangle," which became known as the "Devil's Triangle."

In 1964, following the disappearance of the *Marine Sulphur Queen,* journalist Vincent Gaddis wrote an article for *Argosy* magazine in which he drew together the many mysterious events that had taken place within the triangular area of sea and sky. He called it "The Deadly Bermuda Triangle," thereby coining the famous expression that was to become synonymous with unexplained disappearances the world over. Ten years later, a book by former army intelligence officer Charles Berlitz, simply titled *The Bermuda Triangle,* sold more than twenty million copies and was translated into thirty different languages. In 1976, the book won the Dag Hammarskjöld International Prize for nonfiction and the world became gripped by Triangle fever—and has been ever since. But it is worth noting that even as recently as 1964 the Bermuda Triangle, as we now know it, simply did not exist.

Geographically, the Bermuda Triangle covers an area in the western Atlantic marked by, at its three points, Bermuda, San

Juan in Puerto Rico, and Miami in Florida—although, on closer study of the locations of some ocean disasters attributed to the myth, it would be easy to extend that area halfway round the world. Even the *Mary Celeste*, for example (see page 138), has been connected to the Bermuda Triangle, which would extend the Triangle's boundaries closer to Portugal!

But could there be any truth to the myth—some more prosaic explanation to account for the seemingly paranormal events? Is there anything about the actual geography of the area that might cause so many ships and aircraft to vanish apparently without a trace?

To start with, the sea currents in the area are heavily affected by the warm Gulf Stream that flows in a northeasterly direction from the tip of Florida to Great Britain and northern Europe. The warm current divides the balmy water of the Sargasso Sea and the colder north Atlantic and is the reason the climate in northern Europe is much more moderate than might be expected, considering that Canada and Moscow are as far north as England. Once leaving the Gulf of Mexico, the Gulf Stream current reaches five or six knots in speed, and this affects the heavy shipping in the area in many ways, including navigation.

Inexperienced sailors, particularly in the days before radar and satellite navigation, could very easily find themselves many miles off course after failing to measure the ship's speed with sufficient accuracy, especially when this was calculated by throwing from the bow of the ship a log attached to a rope and timing the appearance of each of a series of knots in the rope as it passed the stern. Failing to do this often enough while sailing in the fast-moving Gulf Stream could quite speedily lead to the crew of a ship becoming hopelessly lost in the vast Atlantic Ocean. Another effect of the fast-moving current would be to scatter the wreckage of lost ships and aircraft over a vast area, many miles from the site of an accident, making it well-nigh impossible for rescue teams to locate survivors.

Then there is the North American continental shelf, which is responsible for the clear blue water of the Caribbean islands. After only a few miles, the shelf gives way to the deepest part of the Atlantic Ocean, an area known as the Puerto Rico Trench. And since it's nearly thirty thousand feet deep, nobody has ever been down there to clear up any mysterious disappearances.

Furthermore, the continental shelf is home to large areas of methane hydrates (methane gases that bubble up through the water after being emitted from the seabed). Eruptions from any of these in the relatively shallow waters cause the sea to bubble and froth, affecting the density of the water and hence the buoyancy of vessels traveling on the surface. Scientific tests have shown that scale models of ships will sink when the density of the water is sufficiently reduced, which could account for the sudden disappearance of various craft within the area. Added to which, any wreckage might be carried away by the Gulf Stream and scattered across the Atlantic in no time at all.

The Bermuda Triangle is also known to be an area of magnetic anomalies, or unusual variations in the earth's magnetic field. Indeed, this area of ocean was once one of the two places on earth where a magnetic compass pointed to true north (determined by the North Star) rather than magnetic north (located near Prince of Wales Island in Canada). The only other place where true north lines up with magnetic north is directly on the other side of the planet, just off the east coast of Japan, an area known by Japanese and Filipino seamen as the Devil's Sea. In both these areas, navigators not allowing for the usual compass variation between true and magnetic north would become hopelessly lost, and mysterious disappearances are equally common in the Devil's Sea. But locals there do not blame UFOs

or sea monsters; they blame human error. Christopher Columbus, the famous fifteenth-century navigator credited with "discovering" the Americas, was one of the first people to recognize the difference between true and magnetic north; and he wasn't at all fazed by the odd compass readings he seemed to be getting as he sailed between Bermuda and Florida more than five hundred years ago.

Magnetic anomalies are also thought to be responsible for the fog that appears to cling to aircraft and boats in the Bermuda Triangle and Devil's Sea. In such cases, the fog gives the strange illusion that it is traveling along with the craft rather than that the vessel is traveling through it, creating a "tunneling" effect for the passengers on board. Many reports have been made of the disorienting effect of this curious fog. In one of the most celebrated instances, the captain of a tug towing a large barge reported that the sea was "coming in from all directions" (because of methane hydrates, no doubt) and that the rope attached to the barge plus the barge itself, only a few yards behind the tug, appeared to have completely vanished, presumably shrouded in magnetic fog.

Another natural phenomenon that might be held responsible for the strange disappearances in the region are hurricanes, notorious in that area of the ocean. These must take their fair share of the blame in bringing down small aircraft and swallowing boats, sending the wreckage to the floor of the Atlantic in minutes and leaving no trace of the craft on the surface.

So what really happened in the case of Flight 19, the USS *Cyclops,* and the *Marine Sulphur Queen?* Let's examine the first of these disappearances in a bit more detail. The squadron leader, Lieutenant Charles Taylor, although an experienced pilot, had recently been transferred to the air station at Fort Lauderdale and was new to the area. Added to which, he was a known party animal and had been out drinking the evening before the fateful day.

A very hungover Taylor tried to find someone else to take over as leader of the training flight—the only point of which was to increase the flying hours of the four apparent novices—but no other pilot would agree to stand in at such short notice. Shortly into the flight, Taylor's compass malfunctioned, and, unfamiliar with the area, he had to rely on landmarks alone. After nothing but open sea, the aircraft eventually flew over a small group of islands Taylor thought he recognized as his home—the Florida Keys.

Flight 19 was in constant touch with flight control and was told to head directly north, which, Taylor thought, would take him straight back to base. But Flight 19 was not in fact over the Florida Keys; it was over the Bahamas—exactly where it should have been. Heading north simply sent the stricken aircraft out into the open Atlantic. Crew members were heard to suggest to each other that they should immediately head west, as their compasses were actually working, but none of the trainees dared to contradict their leader.

With a storm gathering and the sun not visible through the clouds, Taylor refused to listen to his subordinates, accepting the instruction from flight control instead. But when told to switch to the emergency radio channel, Taylor declined, stating that one of his pilots could not tune in to that particular channel and that he did not want to lose contact with him. As a result of this, contact between Flight 19 and Fort Lauderdale became increasingly intermittent.

After an hour of flying due north, and with no land in sight, Taylor reasoned he must be over the Gulf of Mexico, and with that made the right-hand turn, due east, that he thought would bring his team back to the west coast of Florida. But instead, an hour north of the Bahamas and flying over the Atlantic with flight control believing them to be close to the Gulf, this maneuver only served to take them farther out to sea.

Flight 19, miles away from where anybody believed them to be, would then have run out of fuel, ditched into the sea beyond the continental shelf, and been broken within minutes by the storm. The Mariner sent to look for them was, in fact, one of two that were sent to assist. The first arrived back at base safely, but the second exploded shortly after takeoff. (The Mariners, notorious for fuel leaks, were nicknamed "flying gas tanks.") Radio contact had been lost twenty-five minutes into the flight, and debris floating in a slick of spilled oil was found in the exact location where the plane was thought to have come down.

In short, there was nothing mysterious about the accident after all. The official report at first stated that flight leader error was to blame for the loss of Flight 19, but this was then changed to "cause unknown," giving rise to the mystery. Contrary to the fictitious version of events, nobody has ever stated, in an official capacity, that the aircraft simply vanished "as if they had flown off to planet Mars."

The disappearance of the USS *Cyclops* does remain a mystery, however, although heavy seas and hurricanes were reported in the area at the time. It is now thought that a sudden shift in its ten-thousand-ton metal cargo was to blame, causing the ship to capsize with all hands on deck and sink to the bottom of the ocean.

In the case of the SS *Marine Sulphur Queen,* something Triangle enthusiasts rarely mention is that the cargo was made up of fifteen thousand tons of molten sulfur sealed in four giant tanks and kept at a heat of 275 degrees Fahrenheit by two vast boilers connected to the tanks via a complex network of coils and wiring. They also do not tell us that T-2 tankers such as the *Marine Sulphur Queen* had a terrible record for safety during the Second World War and that within the space of just a few years three of them had previously broken in half and sunk. Indeed, a similar sulfur-carrying ship had vanished in 1954 under less mysterious circumstances, having spontaneously exploded before any distress call could be made.

But what clinches it for me is one particular detail: the fact that officers on a banana boat fifteen miles off the coast of Cuba reported a strong acrid odor in the vicinity. The conclusion at the time, but overlooked later by Triangle enthusiasts, was either that leaking sulfur must have quickly overcome the entire crew and a spark then ignited the sulfur cloud, causing a fire that the unconscious crew were unable to put out, or that an explosion had torn through the boat, depositing the crew in the shark- and barracuda-infested waters. Either way, investigators decided the ship must have gone down just over the horizon from the banana boat whose crew had detected the sulfurous odor.

In addition to natural phenomena, there are man-made ones to consider too when it comes to the Bermuda Triangle. The

Caribbean and southern Florida have long been a favorite haunt for pirates, and it's not exactly in their interests to report the ships they've sunk after looting their cargo or the crew they've murdered in the process. Many unexplained disappearances would be far better explained by pirate activity than by extraterrestrial abduction or sea monsters lurking in the deep. The pirates of the Caribbean were not heroes but vicious murderers who took no prisoners and left no evidence of their piracy, and don't let Johnny Depp or Keira Knightley seduce you into thinking otherwise.

The main explanation for the mysterious events of the Bermuda Triangle is sheer invention. There are many examples of writers bending facts to suit their stories (notably in the case of the Loch Ness Monster and the *Mary Celeste*—see pages 121 and 138—or indeed pretty much every story I've covered in this book), which is hardly surprising, since mysterious and ghostly goings-on can be very profitable (as I hope to find out): everyone loves a good mystery.

One of my favorite examples of this is the story of the incident in 1972 of the appropriately named tanker *V. A. Fogg*, which was said to have been found drifting in the Triangle without a single crew member aboard. Everybody had vanished apart from the captain, whose body was found sitting at his desk with a steaming mug of tea in front of him and a haunted look on his face. He had died from shock—or so the story goes.

The truth is rather different, although not lacking in drama. The *V. A. Fogg* had just delivered a cargo of benzene at the Phillips Petroleum depot at Freeport, Texas. As it returned through the Gulf of Mexico with its skeleton crew (and I mean that metaphorically, in case you've still got those Caribbean fellows on your mind) cleaning out the fuel tanks, the ship suddenly exploded and sank. The blast created a ten-thousand-foot-high pall of smoke and, on further investigation, the U.S. Coast Guard found the vessel broken in two on the seabed, one hundred feet below the surface. Their photographic record, including the bodies recovered from the sea, is at complete odds with the story told for the benefit of the Bermuda Triangle mystery, plus, of course, the fact that the Gulf of Mexico is not even in the Bermuda Triangle. I don't mean to be a mystery buster, but we do need to get our facts straight.

To resolve the mystery of the Bermuda Triangle once and for all, I decided to adopt my fail-safe research method of getting to the bottom of things—finding out who has the most money at stake. I don't mean documentary makers, newspapers, or television companies; I'm talking about the insurance industry. Because it is very much in their interests to carry out meticulous research into accidents at sea, we can be fairly certain that they will have looked into any so-called mysteries with considerable care.

Starting with the largest, and oldest, shipping insurance company in the world, Lloyd's of London, we discover that they certainly did take notice of the Bermuda Triangle reports during the early 1970s and issued a statement to *Fate* magazine, published on April 4, 1975. The statement declared that "428 vessels have been reported missing throughout the world since 1955, and it may interest you to know that our intelligence service can find no evidence to support the claim that the 'Bermuda Traingle' has more losses than elsewhere."

So if Lloyd's of London believes there is no mystery to be found in the Bermuda Triangle, then neither should we. But just in case people with minds immeasurably greater than ours are wrong, or even lying to us, then let's do a few calculations of our own. We could start by considering that the surface of the earth is 71 percent water, an area of 13,900,000 square miles. The Bermuda Triangle at its smallest—depending on which author you believe, as some extend the area to cram as many disappearances into their version of the Triangle as possible—is around 500,000 square miles: about 3.6 percent of the world's sea area.

During the last century more than fifty ships, large and small, and twenty aircraft of all shapes and sizes have come to grief in the Bermuda Triangle. If we use those figures and apply the same principle across the planet, we should expect to have lost around two thousand aircraft and boats in total over the past one hundred years, which sounds a little too high. But are twenty accidents per year, small or large, around the world, unreasonable to imagine? Are the events attributed to the Bermuda Triangle any greater in number than they would be in any other section of the ocean of comparable size?

Other mystery makers point to the statistic of one thousand craft lost in the Bermuda Triangle since records began. But they fail to remind readers that records began many centuries ago when Christopher Columbus first sailed west in 1492, which works out to an average of less than two disappearances per year. That sounds about right to me. That, combined with the fact that coast guards have known the reason for the loss of a craft in almost every case—if people would only bother to ask them—should stop the fuss once and for all. This isn't an unusually high percentage of accidents for this area at all in comparison with other parts of the world. The only real surprise is that Lloyd's made any statement at all—if they'd kept quiet, they could have raised their premiums for shipping in that now infamous stretch of sea.

Does Bigfoot Exist?

*What made the oversize tracks found in Bluff Creek,
California, and other parts of America?
Was it a giant ape or just a big jape?*

In 1924, a group of miners working in the Cascade Mountain
Range in the state of Washington were startled to see a huge
simian creature staring at them from behind a tree. Panic-
stricken, one of the men fired at it, and although the bullet appeared to hit the giant ape in the head, the beast ran off,
apparently unharmed. Soon afterward another of the miners,
Fred Beck, spotted it again on the edge of a canyon and again
fired, this time hitting the creature in the back. The group
watched as it fell over the ridge. They scrambled at once down
into the canyon below, but could find no trace of the creature's
body.

However, that evening as it grew dark, the men heard strange
scratching noises outside their log cabin and saw shadowy
gorilla-like faces at the window. The terrified miners barricaded
the door, but soon the creatures were hammering at the roof
and walls. Heavy rocks were thrown and the cabin rocked from
side to side. The men began shooting through the walls in all directions but still the hammering continued, only ending as the
sun rose the next morning. The miners packed up at once and
left the cabin, vowing never to return.

It was only after Eric Shipton famously photographed a giant
footprint on the Menlung Glacier of Mount Everest in 1951,

putting his pickax alongside to show its size, that interest in giant apes began to gather pace. During the 1953 expedition to Everest, when Edmund Hillary and the Sherpa guide Tenzing Norgay were the first to successfully climb the mountain, both men reported seeing oversize footprints. Although Hillary later disputed that these were yeti tracks, there was so much interest in finding out more that the *Daily Mail* sponsored a "Snowman" expedition in the Himalayas the following year. Keen to discover more about America's very own yeti-style legend, John Green tracked down Fred Beck in the late 1960s and interviewed him for his book *On the Track of the Sasquatch,* and the Bigfoot mystery took even firmer root in America.

The word *sasquatch,* applied to the large, hairy hominid in its North American manifestation, was first coined much earlier— in the 1920s—by J. W. Burns. While working as a schoolteacher at the Chehalis Indian Reservation on the Harrison River, he had learned that Native Americans used the words *soos-q'tal* and *sokqueatl* to describe the various "giant men" of their legends. To simplify matters, Burns decided to invent one name to cover all such creatures, and through one of his articles—"Introducing British Columbia's Hairy Giants," published in *MacLean's Magazine* in 1929—"Sasquatch" passed into wider use.

As the public fascination for the giant apeman grew, the media began to report sightings on a regular basis. In 1958, road construction worker Ray Wallace was amazed when his colleague reported finding huge footprints in the dirt at Bluff Creek in northern California, the area they were working in. The local press descended and soon the story was front-page news all over America. Casts were made of the prints, which experts declared genuine. The first newspaper to carry the story, the *Humboldt Times* of Eureka, California, used the name "Bigfoot" in their headline, and the word has since become synonymous with America's favorite mystery creature. When more tracks were found, Sasquatch hunters flocked to the now famous Bluff Creek area to see what else they could discover.

It wasn't until Ray Wallace's death, in December 2002, that the mystery was revealed. Members of Ray's family requested that his obituary should announce that, with his passing, Bigfoot had also died. Ray Wallace immediately became one of the most controversial characters in Bigfoot history when it was revealed that he (along with a handful of his close friends and co-workers) had made the tracks. Investigators soon found out that all of the tracks appeared in areas Ray had worked in. In the early days, that was in Washington State, where the first foot-prints had been found, while more than twenty years later, discoveries were being made farther south, in California. Bigfoot had not been on the move, Ray Wallace had. Family members produced dozens of different oversize foot molds made out of wood or clay that Ray must have spent weeks crafting and hon-ing.

His buddies, by then rather elderly pranksters, showed in television documentaries how they had created the vast footsteps: holding on to a rope tied to the back of a logger's truck being driven very slowly had enabled them to take the giant steps that had so fooled expert analysis. In much the same way as crop-circle makers simply enjoy confounding the experts (see page 66), so did Ray and his pals.

However, despite *The New York Times*'s running the news as a headline story, many Bigfoot researchers have discounted the revelation (not altogether surprising, cynics might say, when their credibility was on the line) and even tried to discredit the Wallace family, threatening them with legal action. One poor haunted soul who spent his adult life in search of Bigfoot evidence wondered why anybody would put so much time into "messing with people's heads." The answer, of course, is because it is fun. Fun, and surprisingly easy.

Nonetheless, a number of scientists and leading members of the Bigfoot Field Researchers Organization (BFRO) are, instead, stating that the footprint molds produced by the pranksters are themselves the fake, not the tracks. In a bizarre piece of reverse logic, some are insisting the Wallace family must prove their claims. John Green, described as one of America's foremost Bigfoot researchers, loftily remarked of Wallace that if he had revealed the footprint mold during his lifetime he "would, of course, [have been] called upon to prove himself." I am unable to see how anybody can become a "foremost researcher" when they have discovered exactly the same amount of genuine evidence of Bigfoot as I have—that is, absolutely nothing.

It was, after all, John Green who interviewed Albert Ostman in 1957 and fell for his tall (in more senses than one) story. Ostman said he had been looking for gold in British Columbia during the gold rush of 1924 when he had been kidnapped by an adult male Sasquatch. The beast gathered up the man in his

sleeping bag and carried him several miles. He was then
dumped on the ground and realized, shortly afterward, that he
was being held by a family of four who would not let him leave
their camp. After six days of captivity, he concluded he was being
considered as future husband material for the young female, so
he fired his rifle into the air, distracting the family for long
enough to make his escape.

When Green asked why Albert had not told his story before,
the aging gold prospector replied that he thought nobody
would have believed him. And few did, except John Green and
his vast fan base of Bigfoot believers ready to leap to his defense
on every issue. But Green did finally concede, in 2007, that he
"would not believe the story if he were told it today."

Take another established piece of "proof"—the footage of a
female Sasquatch filmed by Roger Patterson in Bluff Creek. The
story goes that in October 1967 Patterson and his friend Bob
Gimlin were riding through the area when their horses reared
up and they were both thrown to the ground. As they picked
themselves up, they noticed a "huge, hairy creature walking like
a man" about thirty yards ahead of them. Patterson grabbed
his movie camera and began filming the beast as she loped
away, pausing only once—and looking directly into the camera
lens as she did so—before disappearing from view. The film
has become world famous and has been studied by zoologists,
cryptozoologists, paleontologists, biologists, anthropologists, ar-
chaeologists, and everyone else. And you will be unsurprised to
hear that opinion is divided about whether it is genuine footage
(Bigfootage?) or not.

Leading scientists did, however, conclude at the time that
there was "nothing in the film that leads them, on scientific
grounds, to suspect a hoax." Having now made my own detailed
study of the film, using ultraslow, frame-by-frame-pausing technology obligingly provided by Sony (namely, the DVD player in

my front room), I can now add to the debate. To my albeit un-trained eye, the creature looks suspiciously like a man in a mon-key suit on his way to a fancy-dress party.

Seasoned Bigfoot researchers nevertheless regard the film as a significant piece of evidence, saying that to suggest that it was a hoax would be "demonstrably false"—that old double-negative rhetoric again. But even nonresearchers, including the physical anthropologist Grover Krantz, confirm the film does depict a "genuine unknown creature." Another prominent primate ex-pert, John Napier, is still not entirely convinced but once re-vealed: "I could not see the zipper then and I still can't. Perhaps it was a man dressed up in a monkey costume; if so it was a bril-liantly executed hoax and the unknown perpetrator will take his place with the great hoaxers of the world." So does this mean if he can't see the zip, it can't be a monkey suit? Or had the hoaxer compounded his/her cleverness by employing an early form of Velcro?

In 2004, Greg Long revealed in his book *The Making of Bigfoot* that the grainy clip was in fact an elaborate hoax. Long claims he had managed to trace the monkey suit to costume maker Philip Morris, a gorilla suit specialist from North Carolina. In the book, Morris states he sold the suit to Roger Patterson for $435, and when he saw the Bigfoot photos on the television and in the newspapers a few weeks later, he recognized the suit as the one he had made. Morris claims never to have revealed this infor-mation before because to break "client confidentiality" in such a public manner would have lost him customers. It might have saved millions of research dollars, though.

Greg Long revealed the man in the suit as Bob Heironimus—a friend of Patterson's—who subsequently told *The Washington Post:* "It's time people knew it was a hoax. It is time to let this thing go . . . I have been burdened with this for thirty-six years, seeing the film-clip on television numerous times. Somebody's making lots of money out of this, except for me. But that is not the issue, the issue is that it is finally time to let people know the truth."

John Green, of course, immediately went on the offensive, calling him a liar and declaring Greg Long had made "a fool of himself." And while Heironimus was a known associate of Patterson's and has passed two lie detector tests, and Greg Long has found several independent, but supporting, witnesses, John Green still has yet to provide a single piece of evidence for his case that the film is of a genuine, if as yet unidentified, hairy giant.

Step forward, then, Roger Patterson himself. Unfortunately, he can no longer be called upon, as he died in 1972. However, the other witness to the Bigfoot sighting, Bob Gimlin, is still alive. Bob no longer speaks personally about the film as he is "fed up with the whole Bigfoot thing," but his attorney, Tom Malóne, issued a statement to *The Washington Post* in response to their story about Heironimus's revelation: "I am authorized to tell you that nobody wore a gorilla suit or monkey suit and that Mr. Gimlin's position is that it's absolutely false and untrue." Which seems clear enough, but it is quite possible Gimlin didn't know about Patterson's hoax and was simply used to increase its credibility. Even if he was in on the act, Gimlin has always maintained the film to be genuine, and so any revelation now, forty years after the event, would be somewhat embarrassing for him.

In 1969, another set of tracks was reported—in Bossburg, Washington—that, on closer inspection, revealed that the giant beast's right paw was in fact clubfooted. Experts argued that this indicated that the tracks were very likely to be the first genuine piece of evidence to support the existence of the Sasquatch. Professor John Napier, whose book *Bigfoot* was published in 1973, wrote: "It is difficult to conceive of a hoaxer so subtle, so knowledgeable—and so sick—who would deliberately fake a footprint of this nature. I suppose it is possible but so unlikely I am prepared to discount the idea it is a hoax." Straight from the school of "If I couldn't think of it then neither could anybody else," and with such imaginative minds on the trail of Bigfoot, it is hardly surprising he has managed to elude us for so long.

Despite sightings of Bigfoot reported in every American state except Hawaii and Rhode Island, the creature's natural habitat is said to be the remote woodlands and forests in the Pacific Northwest of America and Canada. The Rocky Mountains have provided many sightings, as have the Great Lakes. But if this is the case, how could he have gotten to Florida and other southern states? The Sasquatch would have had to leave the cover of his remote woodland hideaway, and it is difficult to imagine how such a creature could travel so far without leaving behind at least some credible evidence. You would certainly spot him in the Greyhound bus queue.

But, unfortunately for the wonderfully named Texas Bigfoot Research Center (TBRC), it turns out that most of the evidence found, such as blood or hair samples, footprint casts or photographs, usually turns out to be fake and never, as yet, from an unknown creature. Investigators at TBRC say they receive reports of more than one hundred sightings each year in Texas alone, while on the home page of their website Janet Bord states: "If the skeptics are right and there is no such creature as Bigfoot, then it is a fact that thousands of Americans and Canadians are either prone to hallucinations, or are compulsive liars or unable to recognize bears, deer and vagrants." Quite how tramps became involved is anybody's guess.

Also on the home page of the TBRC website is something that bears further examination. One Rick Noll is quoted stating his reasons why no firm evidence for the existence of a big, hairy, part-man, part-simian-type monster has been found:

1 No one is spending enough time in the woods,
2 Not many know what to do in searching, overlooking things, or vice-versa, seeing things that aren't significant [*sic*] to the task,
3 There are not many of these animals around,
4 They, like most animals in the forest, know how to camouflage themselves quickly and easily,
5 Most encounters with humans are probably mistakes on the part of the Bigfoot, yet researchers are trying to fill in the picture with them as to being something significant.

So there you have it. Five good, solid, scientific reasons why we still have no credible evidence of the existence of Bigfoot. So how is it, then, that despite the use of the whole spectrum of technology—from heat-seeking cameras with night vision to thermal imaging—nobody has confirmed the existence of Bigfoot?

Bigfoot enthusiasts apart, the group of people keenest to obtain as much information as possible of the apeman's existence would be the U.S. government. And as they have surveillance equipment that can detect a small nuclear warhead buried in the desert somewhere near Baghdad, it is fair to assume they would have picked up one of the thousands of Sasquatches that have to exist if all the Americans and Canadians who claim sightings are not lying.

Such a large number of sightings does suggest that Bigfoot, or a relative of his, could well be out there; indeed I, like Janet Bord, refuse to believe that so many people can be lying. But hundreds of small, circumstantial, and unprovable reports do not add up to a single solid fact. It is like pouring thirty separate measures of Jack Daniel's into a large glass. Added together they do not make the drink any stronger in flavor; it still tastes exactly the same. But if you drink it all—as I have discovered through experimentation on your behalf for this very investigation—you will fall over. Scientifically speaking, weak evidence should not become any stronger just because there is lots of it, although it can affect your judgment in the end.

But the Texas Bigfoot Research Center is not the only organization dedicated to finding firm evidence: there are many others throughout America. On December 27, 2003, for example, the Pennsylvania Bigfoot Society (PBS) hosted their fifth annual East Coast Bigfoot Conference (ECBC). The keynote speaker, Stan Gordon—a veteran researcher with over forty years' experience, founder-director of the Pennsylvania Association for the Study of the Unexplained (PASU), and winner in 1978 of the Meritorious Achievement in a UFO Investigation Award— concluded his opening speech linking Bigfoot sightings with known UFO activity in the same areas, although he stopped short of announcing: "Bigfoot is a spaceman." Which I would have done, just for the headline. "There is no doubt the evidence suggests there is something out there," he assured the audience, as they sat there hanging on his every word, then continued: "We just don't know what it is."

Another speaker at the conference, Paul Johnson, a chemistry professor at Duquesne University in Pittsburgh, thought he knew: "Bigfoot is a quantum animal that moves freely between the real world as we know it and a quantum world outside the reach of conventional laws." He went on to explain how, in quantum physics, electrons do not follow the normal rules of physics. Although he admitted his ideas were unconventional, he also noted (contradicting himself in the process) that nothing as large as Bigfoot could behave like an electron in reality, which was a relief because everybody knows that a living being is unable to dematerialize and then reappear in perfect working order in another place. Unless, of course, you are traveling on the starship *Enterprise*, and then you can.

Another speaker at the ECBC, Janice Coy from Monroe County, Tennessee, claimed her family had developed a relationship with a family of Bigfoot (or should that be "Bigfeet"?) since 1947. Her grandfather, having stumbled across an injured Bigfoot, had bandaged its broken leg and allowed it to recover in a barn at the family farm. She claims to have even held a baby Bigfoot in her arms and explained that for years she had tried to obtain photographic evidence, without success. She picked up on Paul Johnson's quantum theory and suggested that was the reason none of her photographs ever showed images any clearer than a "shapeless fuzz." And no one likes to see a shapeless fuzz now, do they?

On one occasion the Sasquatch family, realizing the camera was present on a nearby tripod, used long sticks to retrieve food from a place out of range of the lens. On another, the roll of film Janice submitted to a commercial processing lab was returned to her after the film had been mysteriously overexposed, and every image lost forever. She also claimed she was trying to obtain DNA evidence to provide comprehensive proof of the family's existence; no one asked her why she didn't just pinch a couple of hairs from the baby she had held in her arms. That would have been enough to prove her bizarre claims. But that's enough about the ECBC, so let's move on.

Where DNA testing has been carried out on purported evidence, none has been proved to come from an unknown beast. Usually Bigfoot hairs are found to have come from bison or other common animals. The absence of fossil evidence is another powerful argument that Bigfoot does not exist, although believers counter this by suggesting that the absence of fossil evidence is not evidence of fossil absence, and so it goes on and on and on. But the fact remains that not a single hair, bone, tooth, nail, or claw has ever been found that belongs to a giant hairy manlike being that cannot be explained, and yet there is plenty of evidence found in similar areas that bears, moose, deer, and even dinosaurs and hairy mammoths have left their traces behind them. So why not Sasquatch, if there is one?

The late professor Grover Krantz, a reputable anthropologist, was one of few scientists to state publicly that he believed in Bigfoot. He personally interviewed hundreds of witnesses, studied film footage and photographic evidence, and inspected many plaster casts of footprints and other imprints. He estimated that between two hundred and two thousand Bigfoot lived in the Pacific Northwest of America and he dedicated his life to proving it, but he never turned up any credible evidence that could be regarded as anything approaching proof. However, the professor was unabashed, once suggesting that most animals hide before they die and their bodies are quickly devoured by scavengers, noting that he had "yet to meet anyone who has found the remains of a bear that was not killed by human activity." Which is a fair point, but then he hasn't met everybody yet, has he?

It was Grover Krantz who announced to the world that the clubfooted prints had offered the "first convincing evidence that the animals were real." He also said of other tracks he had studied that a "push-off mound" was "impressive evidence" to him. This was a small mound of soil, present in some Bigfoot tracks, that Krantz had decided was created by the "horizontal push of the front foot just before it leaves the ground." He stated with authority that no artificial rubber or wooden mold would leave such an impression.

More recently, in 2005, a story was told of a young Bigfoot that had been accidentally caught in a bear trap. A boy and his father had taken the beast back home and put him in a cage, but when the Bigfoot became distressed, the boy's father let it go. In a world where everybody now has video cameras, even on their mobile telephones, it is hard to believe that their first instinct wouldn't be to take a close-up picture of the creature. Quite frankly, although this story is reported as genuine, if it turns out to be true, then I will shave my head and become a French monk.

So, in summary, what is still needed is a carcass. That would be ideal, although any Sasquatch fossil or bone would do—just something more convincing than the plaster mold of an oversize footprint made by a carved wooden or plastic shape strapped to the foot of a prankster being pulled along by a slow-moving truck to help create the effect of giant footsteps that "man could not possibly have made." Even the apparently genuine footprints look suspicious to me. Look again at the assessment of the small mound of earth focused upon by the expert Dr. Grover Krantz—caused "by a horizontal push of the front foot just before it leaves the ground." Now go and have a walk across your living room as I just did, and notice how your front foot never leaves the ground, until the other one passes it of course, but by then it is your back foot. So what is he talking about?

As every single apparently credible piece of material evidence of Bigfoot has turned out to be a hoax, then there is nothing else for it—we do need a carcass. Indeed, Krantz himself believed this would be the only way to finally remove any doubt in people's minds as to the existence of Bigfoot, and he called for hunters to bring one in. But that also worries me, because what if the one that is shot turns out to be the only one? Hold your fire after all, fellas . . .

Either way, the search, for some, will continue, and groups of people known by their initials, including Central Ohio Bigfoot Research (COBR) and the Gulf Coast Bigfoot Research Organization (GCBRO), will continue to flourish and attract new members and devise new acronyms. I might even start my own group and call it the Time Wasters And Tricksters Society, of which I am told by some that I am perfectly qualified to be the president. Because if enough people continue to insist Bigfoot, or Sasquatch, is alive and well somewhere in the wilderness, there will always be hoaxers leaving clues for them to find. In reality, it will remain as impossible to prove Bigfoot does not exist as it is to prove you do not have an invisible, silent pink lion standing in your garden, looking at you right now and thinking, "Lunch." You can't prove there isn't one, you know. After all, any absence of evidence for invisible, silent pink lions is not evidence of their absence.

ALBERT JACK
Chief TWAT

The Spine-chilling Tale of the Chase Vault

What terrifying secret is sealed within an old family tomb in Oistins, Barbados?

Nestling in the idyllic range of islands in the Caribbean Sea is the island of Barbados. The most easterly of them, Barbados is also the newest, having been created a mere million years ago when the oceanic plates of the Atlantic and Caribbean collided and a volcanic eruption formed new land in the clear blue sea. First discovered by the Portuguese, who were on their way to Brazil, the island was named Isla de los Barbados ("island of the bearded ones") by the explorer Pedro a Campos after he noted that the fig trees along the coastline gave it a beardlike appearance. The island was first settled in 1511 by the Spanish, who enslaved the natives. But when outbreaks of smallpox and tuberculosis—the European diseases they had brought with them—led to the Caribs dying out completely, the Spaniards abandoned the island. The English then arrived, on May 14, 1625, in the shape of one Captain John Powell, who claimed the land in the name of King James I, and a few years later Captain Henry Powell (no relation) landed with a group of eighty settlers and ten slaves. The island then remained under British rule until its declaration of independence in 1966.

From the seventeenth century onward, the nobles of England who had been awarded land on the island began importing thousands of African slaves to work the newly formed tobacco, sugar, and cotton plantations. Over the next century, Barbados dominated the world's sugar industry and the plantation owners became powerful and successful figures throughout the British empire.

It was one of these landowners, the Honorable Thomas Waldron, who in 1724 built an elegant family burial vault in the cemetery of the parish church in the town of Oistins. It was intended for his married daughter and her family. Seven feet wide and twelve feet deep, and made out of carved coral, the vault was large enough to accommodate the entire Waldron family. The first person to be buried in it was James Elliot, the husband of Elizabeth Waldron. He was also the last of the family to be interred there.

Nobody has since been able to explain why Elizabeth failed to join her husband in his final resting place, nor why the next occupant, Mrs. Thomasina Goddard, was a nonfamily member (unless she was a descendant of the Elliots or the Waldrons by marriage), but what is known is that when the tomb was opened on July 31, 1807, to bury Mrs. Goddard, it was found to be empty. The absence of James Elliot's body was not considered particularly odd at the time, being put down to the work of grave robbers and looters. Rather more unusual was that, soon after Thomasina's death, the Elliot vault passed into the hands of yet another family after being purchased by Colonel Thomas Chase, one of the most hated men on the island.

A plantation owner of unstable mind and volatile temperament, Chase wasn't popular even with his own family. Within a year of the purchase of the vault, tragedy befell the Chase family with the death of the youngest daughter, two-year-old Mary Anna Maria Chase—the result, or so rumor had it, of a fit of violent temper by her father. Nothing, however, was proven, and islanders were left to draw their own conclusions about how the baby had died. On February 22, 1808, the vault was reopened and her tiny lead coffin gently placed on the shelf below the wooden coffin of Thomasina Goddard. Once the funeral was over, Chase ordered his slaves to seal the tomb with a large marble slab set in concrete.

Four years later, on July 6, 1812, the family were back at the crypt for the burial of their teenage daughter, Dorcas Chase, who had died of starvation. While some suggested the young girl had committed suicide to be free of her unpleasant father, others claimed he had locked her in an outbuilding and starved her to death himself. Either way, the marble was cut away and Dorcas's heavy leaden casket was placed alongside that of her sister inside the family vault.

Just over a month later, Thomas Chase himself committed suicide—although there were claims that his slaves had carried out their often repeated threat to murder him. In a land of cruel employers, Chase had been particularly notorious, and there was no shortage of offers to carry his heavy lead coffin, which would have weighed about five hundred pounds, to its final resting place. Presumably people wanted to make sure he had actually gone for good.

Eight slaves carried the casket down the steps of the Chase family vault. As they stepped inside, the men suddenly froze with fear. By the flickering light of their candles they could see that little Mary Anna Maria's coffin was now upside down, standing on end at the opposite side of the chamber from where it had originally been placed. Dorcas's had also moved to the opposite side of the vault, and only Thomasina's coffin remained in its original location. The men inspected the vault and could find no sign of forced entry or any other disturbance. The coffins of the two girls were replaced in their previous positions and their father's casket was settled on the opposite side of the vault. Once the service was over, the men checked for secret passages or other means of entrance before cementing the heavy marble

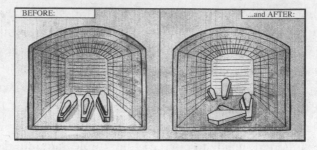

slab back into place, this time using double-strength concrete lest the colonel himself should rise from the dead.

The disturbance was blamed on slaves with a strong grudge against the Chase family. Plantation and slave owners on the islands particularly feared revenge attacks upon their dead, which is why such strong family vaults were built in the first place. In fact, the reverse would have been true: fearing that the evil spirits they called "duppies" might be at work, slaves would stay a long way from cemeteries and graveyards, especially the one housing the Chase tomb.

Four more years passed before the next death, a young Chase relative, Samuel Brewster Ames, who died just before his first birthday. On September 25, 1816, workmen once again broke open the marble seal, but this time they were unable to push open the wooden doors at the vault entrance. A group of the strongest men on the island were called for, and after much effort they managed to force the door open. Thomas Chase's five-hundred-pound lead coffin had been standing on one end with the top resting against the doors, blocking them. The girls had also been disturbed again, while only Thomasina remained peacefully in place.

When the tomb was reopened a month later, for the funeral of the earlier boy's namesake, another Samuel Brewster—killed by slaves during an uprising—it was, once again, in complete disarray, with no obvious signs as to how the disruption had been caused.

The next time the tomb was opened was in 1820 to receive the body of Thomasina Clark, Mrs. Goddard's daughter. By now the mystery of the Chase Vault had spread far and wide, and a crowd of nigh on a thousand curious onlookers were squeezed into the churchyard. The presiding clergyman, the Reverend Thomas Orderson, was accompanied by Viscount Combermere, the governor of Barbados, who was keen to solve the mystery of the disrupted vault, and by island dignitaries such as Major J. Finch, the Honorable Nathan Lucas, Mr. Rowland Cotton (a trusted relative of Combermere's), and Mr. Robert Boucher Clarke. The viscount ordered a thorough inspection of the exterior of the tomb until all present were satisfied it had not been breached. Two masons were then ordered to remove the concrete seal of the marble slab and, accompanied by eight pallbearers, the dignitaries descended the steps.

As the door was pushed open, there was a loud grating sound from inside. This time Dorcas's coffin was found wedged into the doorway. Little Mary Anna Maria's casket had been thrown so violently against the wall it had gashed a chunk from the smooth surface. The other lead caskets had been so chaotically disturbed that Thomasina's wooden coffin appeared to have been smashed in the process and bits of her skeleton lay strewn around the vault.

It was a horrifying sight: some of the slaves fainted while others were violently sick. Combermere and his shocked party were determined to solve the mystery, however. Lady Combermere recorded the subsequent events in her diary:

In my husband's presence, every part of the floor was sounded to ascertain that no subterranean passage or entrance was concealed. It was found to be perfectly firm and

solid and not even a crack was apparent. The walls, when ex-
amined, proved to be perfectly secure. No fracture was visible
and the sides, together with the roof and flooring, presented
a structure so solid as if formed of entire slabs of stone. The
displaced coffins were rearranged, the new tenant of that
dreary abode was deposited and when the mourners retired
with the funeral procession, the floor was sanded with fine
white sand in the presence of Lord Combermere and the as-
sembled crowd. The door was slid into its wonted position
and, with the utmost care, the new mortar was laid on so as to
secure it. When the masons had completed their task, the
Governor made several impressions in the mixture with his
own seal, and many of those others attending added various
private marks in the wet mortar.

Lord Combermere reasoned that anything disturbing the
coffins, even flooding, would leave telltale signs in the layer of
sand on the floor. Then a few months later, a woman who had
been visiting the cemetery reported a loud cracking noise coming from within the Chase Vault, accompanied by an audible
moaning. Her horse became so distressed that it began foaming
at the mouth, later needing sedation. Other horses tethered in
the churchyard broke free and galloped away in fear, straight
into the sea, where they were drowned.

On April 18, 1820, Viscount Combermere and his witnesses
all returned to inspect the vault. The ground had not been disturbed in any way. The seals they had made in the cement remained intact and there was no sign of any foul play. But when
the marble slab was removed and the heavy vault door slowly
pushed open, a scene of complete devastation was revealed.

This time even the lead casket of Dorcas Chase had been
smashed and her bony arm hung out through a gash in the side.
Once again there was no sign of forced entry, or of someone having gained access via a secret passage, nor had the sand scattered

on the floor been disturbed in any way. There were no foot-prints.

Combermere wisely decided to give up trying to solve the mystery, such was the hysteria building up across the island and throughout the empire. This time he ordered that all the bodies be removed and reburied in separate sites in different church-yards. At the same time, a thorough search was made for the coffin of James Elliot, the first inhabitant of the Chase Vault nearly a century earlier, but it was never found. The tomb has re-mained empty ever since.

Later on that evening of April 18, one of the members of the funeral party, Nathan Lucas, was—like Lady Combermere be-fore him—moved to record the events of the afternoon:

> . . . and so I examined the walls, the arch and every part of the Vault, to find every part old and similar. A mason in my pres-ence struck every part of the bottom with his hammer and all was solid. I confess myself at a loss to account for the move-ments of these leaden coffins. Thieves certainly had no hand in it; and as for any practical wit or hoax, too many were req-uisite to be trusted with the secret for it to remain unknown; and as for negroes having anything to do with it, their super-stitious fear of the dead and everything belonging to them precludes any idea of the kind. All I know is that it happened and that I was an eye-witness . . .

Over the following two centuries, much has been made of the events at the Chase Vault: every possible reason has been con-sidered. At first it was thought to have been straight vandalism, such was the dislike among the community of Thomas Chase, but as the heavy coffins would have taken at least six men to move them around, let alone throw them about, and the vault simply wasn't big enough to accommodate that many people, this was ruled out. The absence of footprints or any signs of entry, forced or otherwise, also appears to rule out human interference.

Earthquakes have been considered, especially as Barbados sits on a seismic fault line, but no quakes had been reported during the period in which the vault was disturbed and there was no evidence of any other damage caused, either in nearby vaults or elsewhere on the island. Some prefer the idea that unseen magnetic forces were at work, especially as the coffins were usually found to be facing in the opposite direction to the one in which they were placed, suggesting they had rotated on their own axis. This may also explain why the wooden casket of Thomasina Goddard remained unaffected until it was smashed to pieces by the others. But lead is not a magnetic material. Furthermore, if such forces had been at work, locals would have noticed its effect on other metals in the graveyard, such as iron headstones or steel plaques. The church bell would surely have kept ringing too.

The wildest theory about what had caused the disturbances in the Chase Vault actually came from the creator of Sherlock Holmes, Sir Arthur Conan Doyle, who, maybe unsurprisingly, appears to crop up in a number of mystery stories (including two in this book—"Fairies at the Bottom of the Garden," page 101, and "Whatever Happened to the Crew of the *Mary Celeste*?" page 138). Conan Doyle believed supernatural forces had been at work but was unable to offer any further explanation except to suggest that the coffins had been moved by the spirits of the two family members who had apparently committed suicide and were therefore "cursed and restless" and in conflict with each other. Indeed, since Dorcas and her father have been separated, there have been no other signs of disturbance at any of the new grave locations.

Gas emitted from the decomposing bodies was considered but soon ruled out as incapable of disturbing a heavy lead coffin.

The only other suggestion that comes close to fitting the facts would be a flood. Natural flooding of an underground vault would disrupt the coffins, causing them to float around and come to rest in a different place as the water subsided. But that wouldn't explain why the coffins were standing on end; nor was there any evidence of water damage each time the vault was reopened. It seems that the mysteries of the Chase Vault have never been adequately explained, and probably never will be. I think we're going to have to mark this one "unsolved."

The Real-life Agatha Christie Mystery

How did the world's favorite crime writer become involved in a mystery of her very own?

Agatha Miller was born in 1890, the youngest child of a wealthy American businessman in England. But after her father contracted double pneumonia, he was unable to provide for his young family and sank into a depression, dying when Agatha was only eleven. The poverty-stricken Millers almost lost their home as a result. The lesson was a harsh one for the young Agatha, and her continuing sense of financial insecurity was later to have disastrous consequences.

At a dance in Devon in 1912, Agatha, now an attractive twenty-two-year-old, met a tall, dashing young army officer. Archibald Christie had trained at the Royal Woolwich Military Academy in London and had been posted to Exeter soon after he had been commissioned. Over the next two years, they slowly fell in love. When war broke out in 1914, Archie was sent to France. During his first return on leave later that year, the couple quickly got married. While Archie served in Europe, Agatha became a voluntary nurse at the Red Cross Hospital in Torquay and spent her many free hours (not many casualties were sent to Torquay) reading hundreds of detective stories.

She was desperate to be a writer like her elder sister, Madge, whom she idolized and whose stories were regularly published in *Vanity Fair*. In a moment of inspiration Madge challenged her to write a good detective story, Agatha's favorite genre. At the time, Torquay was full of Belgian refugees, and her first story featured a Belgian detective—one Hercule Poirot—who would become one of the most popular fictional detective characters in the world.

After the war ended, Archie started work at the Air Ministry in London, and the couple had a daughter in 1919. The Christies were struggling to make ends meet and so Agatha decided to approach a publisher with *The Mysterious Affair at Styles,* her first novel. John Lane at Bodley Head read and liked it. He persuaded the inexperienced young writer to sign a five-book deal with them, heavily weighted in their favor. She grew to regret this, however, when despite the book's success and sales of two thousand copies in America and Great Britain, she received only £25 in royalties.

Her final book for Bodley Head, *The Murder of Roger Ackroyd* (1926), had a controversial twist—the book's narrator turned out to be the murderer—and it received lots of attention in the press as a result. That same year, Agatha changed publishers. Collins offered her an advance of £200 for her first book, an impressive sum in the postwar 1920s.

The Christies moved into a new house in Berkshire that she called Styles, after her first novel. Flushed with her growing success and sudden minor celebrity status, Agatha failed to notice her husband's increasing resentment at her refusal to share any of her new income with him. Despite the fact that they were now comfortably off, she insisted on careful economy and thrift, something clearly related to her father's previous loss of wealth. Unknown to Agatha, Archie now began to spend a lot of time with Nancy Neele, a secretary ten years Agatha's junior, whom he had met on the golf course.

But as her financial situation improved, other aspects of her life took a further turn for the worse. In April that same year,

Agatha, en route to visit her mother in Torquay, felt a strong premonition that her mother was dead. Then, upon her arrival in Torquay, she was informed that her beloved mother had, in fact, died suddenly and unexpectedly, from bronchitis. Later that year, returning from a foreign holiday, Agatha got wind of her husband's adultery. She immediately confronted Archie and collapsed in shock when he admitted that he had indeed been having an affair for the previous eighteen months. Agatha begged Archie to stay so that they could try to save their marriage, but Archie refused, moving out of the family home and into his club in London.

Then, on the morning of December 4, a cold and wintry day, the Surrey police were called to the scene of a motor accident at Newlands Corner in Guildford. Agatha Christie's car had been found halfway down a bank and partly buried in some bushes. The headlights were blazing, a suitcase and coat had been left on the backseat, but there was no sign of the author. Upon discovering that the police suspected either suicide or murder, the press descended on Guildford and the Christies' Berkshire home, thrilled at the prospect of a real-life mystery. By the following morning, the disappearance of the still relatively little-known author was a front-page story on every national newspaper. Agatha Christie was suddenly big news.

In one of the finest publicity coups of all time (intentional or otherwise, but for her publisher the cheapest), members of the public were offered rewards for sightings, and newspapers reveled in their ongoing real-life whodunit, with new "evidence" regularly being reported. Some observers suggested that it must have been Archie—with much to gain from his wife's death—who had been responsible for her disappearance. But then it was discovered he had been at a weekend party with his mistress. The focus then moved on to Nancy Neele, and she was hounded by the press, eager to find a culprit. For ten days Surrey police combed the area for evidence, and reports of sightings continued to pour in. People scoured her books for clues (the police actually dredged a pool that was featured in one of Agatha Christie's books and in which one of her characters had drowned) and followed the story avidly in the newspapers.

The breakthrough finally came when, after ten days, the headwaiter at the Hydropathic Hotel (now the Old Swan Hotel) in Harrogate, Yorkshire, realized that the mysterious novelist he had been reading about for nearly two weeks looked exactly like a stylish female guest who had checked in under the name of Mrs. Neele, claiming to be from South Africa. For ten days "Mrs. Neele" had been singing, dancing, and enjoying the company of the other guests while, like them, also following the Agatha Christie mystery in the newspapers.

The police were called and Archie Christie traveled to Harrogate to identify his wife. In a scene that could have come straight from a Christie novel, Archie placed himself at a table in the corner of the dining room, hidden behind a large newspaper. From there he watched his wife enter the room, pick up the papers containing her picture and the story of the continued search, and sit at another table. The hotel manager later said that as Archie Christie approached his wife, she "looked distant as though she recognized him but could not remember where from."

So as the police were scouring the hills around Guildford on their hands and knees, Agatha had been alive and well up in Yorkshire rather than lying dead at the bottom of a lake somewhere in Surrey. Needless to say, the police were not impressed; indeed, some newspapers claimed the whole thing had been a publicity stunt. The press pack raced to Harrogate nevertheless, but few believed Archie when he informed them that Agatha was suffering from memory loss. There was a public backlash with demands for the police to be repaid the estimated £3,000 cost of

the search for the missing novelist—indeed, Guildford residents blamed the next increase in their taxes on her. Reviews of her next book, *The Big Four,* were spiteful as a result, but Agatha Christie was now nationally famous and sales of this new work topped nine thousand copies. The whole affair was a marketing man's dream, with all of Agatha's earlier books being reprinted and enjoying healthy sales.

But the personal outcome for the author was not so positive, as Archie promptly divorced her and married Nancy Neele. In 1930, Agatha met and married the archaeologist Max Mallowan, with whom, having learned her lesson, she immediately shared her resources. None of the parties involved ever spoke of the writer's mysterious disappearance again.

But the debate continued. Could Agatha Christie have had a nervous breakdown? After all, how could she have read about her disappearance in the newspapers and not even recognized a picture or description of herself? For that matter, how could the other guests not have recognized her earlier? Many commentators have suspected a conspiracy—a pact of silence between the writer and her fellow guests.

It was only after the death of Agatha Christie, in January 1976, that the mystery was finally unraveled. It is obvious from the details that the whole affair was in fact far from a publicity stunt. Indeed, Agatha was mortified at seeing so much made of her disappearance. The great mystery of the 1920s, involving the crime writer who was to become one of the most famous and successful in the world, is in reality an easy one to solve.

In 1926, as we have seen, Agatha Christie's world was thrown into turmoil by the sudden death of her mother and the breakdown of her marriage. The mixture of grief, anger, and humiliation she felt following these events led Agatha to the verge of a nervous breakdown, and for the first time in her life she began to behave irrationally. On the morning of Friday, December 3, Agatha and Archie had a major argument about Archie's intention to spend the weekend in Surrey at the house of a friend. He didn't want her to accompany him because, as the writer later discovered, Nancy Neele was going to be present. Such a public breakdown of her marriage was incredibly humiliating, and so— fueled by despair, vengeance, and plain old attention seeking— Agatha, assisted by her sister-in-law Nan, hatched a plot worthy of one of her own novels.

At 10 P.M. on December 3, after Archie had left for the weekend, Agatha drove to Newlands Corner, parked on the edge of the road, and pushed her car down the bank, leaving a suitcase and coat on the backseat and the headlights on, presumably to ensure the car would be discovered. Carrying a second suitcase, she then walked or received a lift to West Clandon station nearby, from where she caught the train to London. After staying the night with Nan, she wrote a letter to Archie's brother Campbell and posted it at 9:45 A.M. on the Saturday, informing him she was traveling to the hotel in Harrogate. She addressed the letter to his office, knowing it would not arrive until at least Monday morning. In the meantime, she was fully expecting the car accident to ruin Archie's weekend, and that of the other guests, who, she presumed, would all be out looking for her rather than having fun without her. When Campbell received the letter on Monday morning, she thought that everything would then die down, and she herself, no doubt, already had her own story worked out about how she could explain the events to her own advantage and to Archie's further misery.

Unfortunately, when Campbell opened the letter that Monday, he hardly looked at it and then managed to lose it, leaving Agatha's whereabouts unknown and the so-called mystery in the hands of the frenzied press. Agatha, clearly alarmed that her mind games had rapidly become so public and out of her control, decided to lie low to consider her next move. Perhaps she would have continued to hide—clearly she hadn't expected anybody to recognize her; or perhaps she would have fled abroad to escape the growing scandal.

It is intriguing to think what Agatha's next real-life storyline would have been if the headwaiter at the Harrogate Hydropathic Hotel had not finally recognized the author. But let us be grateful that he did, because some very fine stories subsequently began to flow out of this now famous author. I am off now to leave my car at Beachy Head to see how many of you come looking for me. If, after a week or so, nobody has tracked me down, try the Old Swan at Harrogate. I don't want to be left there too long.

Committing the Perfect Crime: The Mysterious D. B. Cooper

··

What happened to the famous hijacker who jumped off a plane carrying a small fortune?

The offense on the face of it was a simple one, but the mystery surrounding its aftermath has passed into legend. On November 24, 1971, a man going by the name of D. ("Dan") B. Cooper hijacked a Boeing 727 on a domestic flight and demanded $200,000 from its owners, Northwest Orient. Confident they would catch the hijacker, the company agreed to pay the cash in exchange for their passengers.

But the hijacker had other plans. After the aircraft had taken off again, minus its passengers and with D. B. Cooper $200,000 richer, he strapped himself to a parachute and jumped out into the cold night. He was never seen or heard of again, so if he survived the jump, it had been the perfect crime. If not, of course, he had been the perfect idiot. Either way, D. B. Cooper became an instant celebrity among the tie-dyed, hash-smoking hippies of the early 1970s, when hijacking had rather more of a romantic/revolutionary feel about it than it does today when terrorists are suspected at every turn. Despite one of the biggest manhunts in American history, including amateur investigations, books, TV documentaries, and films, nothing more is known about D. B. Cooper today than was known on the day of his daring airborne stunt.

So let's look at the events in a bit more detail. At 4 P.M. on that particular day in 1971—the fourth Wednesday in November, Thanksgiving Eve—a soberly dressed businessman approached the counter of Northwest Orient Airlines at Portland International Airport and bought a one-way ticket to Seattle for twenty dollars. The businessman, who gave his name as D. B. Cooper, was allocated seat 18C on Flight 305, which left on time at 4:35 P.M., climbing into the cold, rainy night with thirty-seven passengers and five flight crew on board.

Shortly after takeoff, the passenger sitting in seat 18C beckoned to an attractive young stewardess, Florence Schaffner, and passed her a note. This was such a common occurrence between businessmen and the flight crew that Schaffner, believing

Cooper had given her his phone number, simply smiled and placed it, unread, in her pocket. The next time she passed seat 18C, Cooper whispered, "Miss, you had better read that note. I have a bomb." She duly read the note and rushed to the cockpit to show Captain William Scott. The captain then instructed Schaffner to walk to the back of the plane and, so as not to alarm the other passengers, quietly sit next to Cooper and try and gather more information. As she sat down, the hijacker opened his briefcase and wordlessly revealed a device consisting of two cylinders surrounded by wires. It certainly looked like a bomb to the young stewardess.

Captain Scott then radioed air traffic control with Cooper's demand of $200,000 in used bills, together with four parachutes, two for him and the others for two of the crew he intended to take with him as hostages. The FBI was alerted, and they ordered Northwest Orient's president, Donald Nyrop, to comply fully with Cooper's demands. After all, they reasoned, where was he going to go? No one could jump from a jet airliner and survive. There was also the safety of the other passengers to consider, together with the negative publicity such a hijacking would generate if the company refused to comply; Nyrop felt $200,000 was a small sum to pay under the circumstances. Cooper then instructed the pilot to stay in the air until the money and parachutes were ready, and soon heard Captain Scott announce to his passengers that a small mechanical problem would require the jet to circle before landing. The rest of the passengers remained unaware of the hijacking and Flight 305 finally landed at 5:45 P.M. at its intended destination.

Once Cooper was satisfied that the money, all in used twenty-dollar bills, and the parachutes had been delivered, he allowed the passengers to leave. At 7:45 P.M., with only the pilot, copilot, one flight attendant, and himself remaining on board, Cooper told Captain Scott to fly toward Mexico. He instructed him to fly at a low altitude of ten thousand feet (instead of the usual thirty thousand feet), and with the landing gear down and the wing flaps set at 15 degrees, thus indicating a detailed knowledge of flying. Unknown to him, however, the plane was being closely tracked by two United States Air Force F-106 jet fighters, using a state-of-the-art radar detection system.

As the flight crossed southwest Washington, Cooper ordered the pilot to slow his speed to 150 knots and the rest of the crew to remain at the front of the plane with the curtains closed. At 8:11 P.M. the rear door warning light came on, and this was the last anyone saw of the mysterious D. B. Cooper. Even the air force pilots shadowing Flight 305 in their jet fighters failed to see him jump.

After landing safely at Reno airport, the agreed-upon destination, the crew waited in the cockpit for ten minutes for further instructions. None came, and air traffic control also confirmed they had not received any instructions from Cooper. Cautiously Captain Scott called the hijacker over the intercom and, on receiving no response, nervously opened the cockpit door. Cooper had vanished, having taken everything with him, including his briefcase bomb, the canvas bag full of twenty-dollar bills, and his hat and coat. All that remained were the three unused parachutes. Cooper had done the unthinkable. He had jumped out of a commercial passenger jet and into the cold, wet night, thousands of feet above the ground. He had completely disappeared, never to be seen again. Nobody could prove he had survived and therefore got away with his crime, but, as even the FBI admitted, nobody could prove he was dead either.

The FBI calculated that the likely landing area for the skydiving hijacker was southwest of the town of Ariel, close to Lake Merwin, thirty miles north of Portland, Oregon. The eighteen-day manhunt that followed failed to reveal a single trace of the hijacker, but then all the FBI had was a description of a fit, six-foot, olive-skinned man of Mediterranean appearance, clean-shaven and wearing a dark suit, which narrowed the search right down to about a billion people, worldwide. They had some work to do.

It was soon apparent to the authorities that they were dealing with a meticulously planned crime, well thought out in every detail. First of all, Cooper had had no intention of taking any hostages with him: his request for four parachutes was simply to ensure that no dummy parachutes were delivered. Cooper had also worked out the weight of the ten thousand twenty-dollar bills as twenty-one pounds. If he'd asked for smaller denominations, they would have weighed considerably more and created a risk when landing, while larger denominations would be harder to pass on, thereby creating a risk of being caught. Hence twenty-dollar bills were perfect for Cooper's purposes.

He also knew that the Boeing 727–100 has three engines, one high on the fuselage immediately in front of the vertical tail fin and two others on either side of the fuselage just above the horizontal tail fins. This meant that neither the engine exhausts nor the intakes would get in the way when he lowered the rear steps and threw himself out into the night, which led to speculation he had targeted Flight 305 specifically for its engine layout.

Cooper also insisted that the pilot not pressurize the cabin, knowing he would be able to breathe naturally at 10,000 feet (but no higher) but reducing the risk of air rush as the door was opened. And as he was fully aware of the 727's minimum flight speed with a full load of fuel, as well as the wing-flap settings required, and appeared to know that the 1,600 mph F-106 fighters would no longer be able to escort the jet once the aircraft speed had reduced to around 150 knots, this gave Cooper the window of time he needed to jump unseen, suggesting to many he was either a serving or retired airman.

The only mistake he made was to leave behind eight Raleigh cigarette stubs and his tie and tiepin, but even this evidence has led the police nowhere. There were also sixty-six fingerprints on the plane that could not be matched to the flight crew or any of the other passengers. Considering the number of people traveling on a commercial airliner in the course of a few weeks, this was regarded as unreliable evidence, although an exhaustive check with FBI records revealed no match anyway and D. B. Cooper's real identity remained unknown. That he could recognize McChord Air Force Base as the Boeing 727 circled Seattle-Tacoma airport also provided a clue, as did his lack of a regional accent observed by the ticket agent who allocated his seat. This all led FBI investigators to conclude Cooper was local and with a background in either military or civil aviation, possibly from McChord Air Force Base itself.

Appalling weather the day after the hijacking interrupted the search through the vast wooded area Cooper had probably landed in. But the full-scale land and air search that took place over the ensuing weeks revealed no trace of Cooper, the distinctive red and yellow parachute, or, most important, the cash. The police search team did discover the body of a missing teenager, but Cooper himself had vanished, which seems to disprove the theory that he had been killed during the jump or on landing. The FBI even checked the national database for any criminal by the name of Dan Cooper, or D. B. Cooper, in order to find out if, on the off chance, this otherwise meticulous and thorough hijacker had been stupid enough to buy a ticket in his own name. He wasn't, although a genuine Dan Cooper in Portland did undergo an uncomfortable few hours of questioning before being released without charge.

The FBI circulated a wanted poster throughout the States, with an artist's impression of Cooper based upon eyewitness accounts, but it could have been a picture of just about any average

American on the street, and as many as ten thousand false sightings were reported. As it was, the FBI interviewed more than fourteen hundred people, but to no avail. The story held the popular imagination for a long time, the newspapers ridiculing the unsuccessful FBI investigation in the process. Eventually the hijacker, named as "John Doe, a.k.a. Dan B. Cooper," was charged, in his absence, with air piracy at a federal court in 1976.

The American public, on the other hand, was in the process of elevating D. B. Cooper to the status of a legend as the mystery around him continued to grow. Bars in the area of Ariel and Lake Merwin set up D. B. Cooper shrines, which remain to this day, and hold D. B. Cooper "days," with local parachute clubs even reenacting the jump on the day before Thanksgiving every year.

That is what we all like most about this sort of history. Nobody was hurt, it involved extraordinary courage, and nothing has been found since. Not even Cooper's hat, coat, or briefcase. And that is why we all want Cooper to still be alive, and not to have been lying at the bottom of Lake Merwin all these years. We like the idea of Cooper jumping out of a passenger jet with the loot, landing and then dusting himself off, picking up his briefcase, putting on his hat, pausing only to straighten its brim, and being back in the office by nine.

But the FBI does not share our warmth toward the mystery man. Agent Ralph Himmelsbach spent eight years at the head of the investigation and was unable to hide his bitterness, calling Cooper a "dirty rotten crook," a "rodent," and nothing more than a "sleazy, rotten criminal who threatened the lives of more than forty people for money," oh—and "a bastard."

Himmelsbach once snapped at a journalist who inquired about Cooper's growing status as a hero. "That's not heroic," he shouted. "It is selfish, dangerous, and antisocial. I have no admiration for him at all. He is not admirable. He is just stupid and greedy." Himmelsbach retired from the FBI in 1980, his work incomplete, to run his own charm school in the Deep South. In his subsequent book, *Norjak: The Investigation of D. B. Cooper*, Himmelsbach tried to promote what is known as the "splatter" theory, meaning Cooper had been killed as he hit the ground. This is dismissed by most, as the body, highlighted by its bright red and yellow parachute, would have turned up sooner or later. When pressed by reporters about why the body had not been found despite a legion of police, the Army Reserve, volunteers, and Boy Scouts all searching, Himmelsbach surprised everybody, including, I imagine, the FBI, when he insisted they had all been looking in the wrong area all the time, despite the Feds' reenacting the jump in an effort to pinpoint Cooper's drop zone.

In 1980, an eight-year-old boy was playing by the river and discovered a bag of cash totaling $5,800, all in twenty-dollar bills. His father, aware of the D. B. Cooper mystery, immediately took the cash to the police, who checked the serial numbers and confirmed this was part of the missing money. Hopes of a conclusion faltered on discovering the cash was found nearly forty miles upstream of where the police now believed Cooper to have landed. This was compounded by the geologists who claimed, having studied the notes and assessed their rate of deterioration, that the money must have been placed in the water in about 1974, three years after the hijacking. Despite these discrepancies, Himmelsbach considered this evidence proof of his splatter theory. He claimed Cooper must have landed in the lake on that dark night and drowned. But the resulting search by scuba divers with modern sonar equipment failed to find any further clues.

Few people outside the FBI believe this theory. Instead, many believe Cooper's careful plan included dropping a few bags of money at a later date to serve as a red herring. It would appear that Cooper had thought of everything, which is why he is probably still at large. The hijacker had a further stroke of luck when on May 18, 1980, the volcano near the site of his purported landing, Mount St. Helens, erupted with such force that the landscape was changed forever, no doubt concealing many undiscovered clues. But there is, however, one more important piece of evidence for us to consider.

In 1972, an embarrassed FBI produced a thirty-four-page booklet detailing the crime and, more important, including photographs of the money and listing every single serial number of the ten thousand bills. The booklet was sent to every bank and financial institution in America, with copies to the national media. However, despite rewards on offer of up to $150,000 for the production of even one solitary bill, none have ever turned up in the American system (with the exception, that is, of the $5,800 discovered in the water). Like Cooper, they have simply vanished.

But this fact alone does not mean Cooper is dead, as most countries around the world, especially developing nations, trade in dollars, and so the money could have turned up anywhere. Even so, the police expected at least one bill to have been found over the years, and that leaves investigators even more baffled. For nothing to have been seen or heard of Cooper, dead or alive, nor for a single banknote to have reappeared, is hard to imagine. And yet this quite literal vanishing into thin air is exactly what did happen.

The problem about carrying out the perfect crime is that then everyone else wants to try it too. The following year produced no fewer than four copycat jumps, and although one, the first effort, did end in a splatter landing, the following three hijackers all landed safely but were arrested at the scene or soon afterward. But then there was a new and interesting development. On April 7, 1972, four months after the Cooper hijack, a

man checked in as James Johnson on United Flight 855 traveling from Denver to Los Angeles. Just after takeoff, Johnson put
on a wig, fake mustache, and sunglasses and gave the stewardess
a note. This read:

> Land at San Francisco International Airport and taxi to Run-
> way 19 Left [a remote part of the airport].
> Send for a refueling truck, but no other vehicles must ap-
> proach without permission.
> Direct United Airlines to provide four parachutes and a
> ransom of $500,000 in cash.

The captain carefully followed the instructions and the aircraft
was soon back in the air again, this time heading for Provo,
Utah. After an hour and a half, Johnson instructed the captain
to reduce altitude and speed and depressurize the cabin, in a
carbon copy of Cooper's plan. Except that a co-pilot glanced
around the cockpit door just in time to see Johnson expertly slip
on the parachute, open the rear door and jump.

The FBI started their investigation the minute the aircraft
landed at Provo. This time they had a cast-iron clue. Johnson
had left a single clear fingerprint on an in-flight magazine. They
were initially baffled, as Johnson had no criminal record and no
match was found for the print. But then they had a breakthrough. In a telephone call to the FBI in Salt Lake City, a young
man gave the police the plan of the hijacking, including details
not yet made public.

He claimed his friend Richard McCoy, Jr., had boasted about
the plan to him, and the preparatory details he described were
in fact identical in every way to those of the hijacking of Flight
855. McCoy was twenty-nine years old, married with two young
children and studying police science at Brigham Young University. He was also a Vietnam veteran, former Green Beret heli-

copter pilot, and specialist paratrooper. The FBI checked his service fingerprint record and found an exact match to the print found on the plane. The handwriting on the ransom note also matched McCoy's samples in his military file. This time they had their man.

Two days later, McCoy was arrested at his home, where police found a parachute suit and a bag of cash containing $499,970. The FBI asked the trial judge to make an example of McCoy to deter further copycat hijackings, and the young man received a sentence of forty-five years without parole. But within months he had escaped from prison. He was eventually tracked to a house in Virginia Beach, where he was shot dead during the ensuing gun battle to rearrest him.

The similarities between the two crimes, in particular the evident flying expertise in each case, led to speculation that McCoy himself was in fact D. B. Cooper, and certainly the tie left behind by Cooper was similar to McCoy's Brigham Young University tie. It seems pretty unlikely, though: How would the D. B. Cooper money get into the river two years after McCoy's death, for instance? Although it might explain why no money ever reentered the system, as McCoy may have stashed it away for the future, and it has remained hidden ever since.

The truth is that the identity of D. B. Cooper remains a mystery, and each year the American media remind the public by way of anniversary articles and features, although to this day nobody has ever produced a credible theory, backed up with indisputable evidence, as to the identity or whereabouts of either Cooper or the money.

Who Really Makes Crop Circles?

....

*Strange formations in fields of wheat and other crops
have been appearing since the 1970s. Are they made by
aliens parking their spaceships, or is the explanation
rather more down-to-earth?*

Of all the subjects I've explored for this book, the one I was most looking forward to finding out more about was crop circles. For years I had been hoping there would be an extraordinary paranormal explanation for crop-circle appearances or, better still, that we were being visited by aliens from other worlds. Then, when it became clear that most of the circles were in fact hoaxes, I relished the thought that I would finally have the chance to dismiss all the crop-circle fanatics I have heard on the radio over the years talking with great passion about the temporary parking of invisible spacecraft in a field, leaving behind an imprint in the flattened wheat when they zoom back into outer space. The only extra bit of evidence offered for this startling conclusion appears to be that witnesses claim flocks of birds veer around crop-circle sites as if avoiding or circumnavigating something the rest of us cannot see.

I was looking forward to poking fun at the gaping holes left in the arguments of these so-called experts expounding their elaborate theories on the six o'clock news. Indeed, I wish I had phoned in once, just once, and asked that sanctimonious old fellow on the BBC, who appeared to be contradicting himself every five minutes, a couple of simple questions: How, then, do you explain the crop circles that have appeared directly underneath power lines—how could a spaceship have landed there? And why choose that spot when there was plenty of field in which to land without being so inconvenienced? In fact, I would pose these questions to him now if I hadn't switched the radio off through sheer boredom, not bothering to find out his name.

At the same time I hoped that the circles weren't all hoaxes and that I could also expose the hoaxers, who seemed to be making a real nuisance of themselves. But having studied the evidence and looked at the extreme lengths to which the hoaxers have been prepared to go, I am no longer so sure I can.

The very first recorded example of a crop circle is a woodcut dated 1678 and entitled "The Mowing-Devil: or, Strange News out of Hartford-shire" [*sic*], and it shows a devil cutting out a pattern in a field of wheat using a scythe. The text explains that the greedy farmer had refused to pay a reasonable fee to the workers for the harvest that year and announced he would rather have the devil do the work instead. The following morning he woke to find Old Nick had done just that and the farmer became too scared to enter the field ever again, and so the harvest, and his occupation, were lost.

However, crop circles didn't come to the public's attention until three centuries after this—in the late 1970s. Their relatively sudden appearance caused a sensation as the British press fought each other for the best photographs. Keen to get ahead of the trend, several of the newspapers began employing circle makers to create ever more elaborate patterns throughout the countryside and, in the process, obviously guaranteeing themselves the first and best pictures of this mysterious new phenomenon sweeping the land.

But while newspapers may have employed circle makers to create some of the circles they photographed, this wasn't known about at the time, and the circles appeared, for the most part, not to be man-made in origin. During the early 1980s, crop circles began to attract serious interest with the discovery of a formation at a place called Cheesefoot Head, a high point on the chalk downlands close to Winchester. Here, a circle fifty-two feet in diameter was flanked by two others exactly half the size symmetrically placed north and south of the larger circle. The perfect spiraling of the flattened straw and precision placement of the formation proved very difficult to explain at the time, scientifically or otherwise. Other formations soon began to appear in wheat, grass, rape, and many other crops. Researchers started to study the designs and began to believe that individually, or collectively, they amounted to coded messages or directions to something as yet unknown, by intelligent beings also as yet unknown. Other physical features were pointed out, such as the way a circle was aligned with the sun or moon, leading to the idea that supernatural forces were at work across the English countryside. It seemed that such mathematically precise formations—especially the increasingly complicated circles that were emerging by the late 1980s—were not, and could never be, the work of man himself.

Then in 1991 two elderly landscape artists, Dave Chorley and Doug Bower, fired an arrow directly into the heart of the ever-growing crop-circle fan club by admitting they had been making them by hand since the 1970s, after reading about the famous so-called "saucer nests" (impressions left in the crops by some disc-shaped object) that had appeared in Australia. By the time they retired from the "business," Doug and Dave had successfully completed over 250 crop circles across southern England.

Doug and Dave's inspiration, the Australian saucer nests, were a collection of seven circles in the oat fields near Bordertown, Australia. Predating these, and providing the original source of inspiration for all the modern-day crop circles that followed, was the Tully Saucer Nest, which appeared on January 19,

1966, on a piece of land at Horseshoe Lagoon near Tully, in northern Queensland.

After the fruits of Doug and Dave's labors in Hampshire and Wiltshire started to attract public attention, a veritable epidemic began to spread. Crop circles popped up everywhere. Each was immediately studied by experts, who either dismissed it as a hoax or accepted it as genuine—that is, not created by man. Artists competed to confound these so-called experts: having one of their circles "proved" genuine was the highest accolade— albeit hugely ironic that for a circle to be good enough to be considered genuine implied that no artist had been involved in its creation. And this was a form of art, after all, even if the medium was rather avant-garde: during the 1980s and 1990s, it was on a par with cutting a cow in half, preserving it in formalde- hyde, and displaying it at the Tate Modern. But it's all art, isn't it?

Not according to the scientists and followers of the growing crop-circle community, it wasn't; it was deception, and it contin- ually interrupted their serious research into what appeared to be a brand-new type of paranormal activity. As their resentment grew, so the work of the circle makers became increasingly elab- orate. As soon as one design was considered too intricate for hu- mans to create, another one would appear that was even more complicated. And the (probably cashing-in) stories from farm- ers about strange lights and howling animals deepened the mystery, as did reports of military helicopters apparently hover- ing around circle sites.

One crop circle appeared in a field close to Stonehenge, the spiritual home of the Druids, shortly before the summer sol- stice in 1996. Predictably enough, the summer solstice, falling on or around June 21, is Stonehenge's busiest day of the year, and in 1996 it was even busier. The great unwashed descended in force to appreciate the new mysterious formation and spent ages analyzing each bent stem of wheat, taking electromagnetic recordings of earth samples, and recording detailed cross-measurements of the circle.

The press arrived in their helicopters, and photographs of the circle appeared on television and in every newspaper. When crop-circle enthusiasts began to turn up in their anorak-clad droves, the farmer erected a shed and charged an entry fee to his field, to help "compensate for the damage." By the end it is estimated he had collected over £30,000, a good day's work when set against the £150 of damaged wheat. "That will be treble gin and tonics all round, please, landlord, and keep an eye out for the tax man."

Crop-circling has become big business: small companies offer gullible businessmen helicopter flights over crop-circle formations. Bus tours are provided, hotels are always full in the vicinity of new designs, and local tradespeople benefit from the arrival of enthusiasts. Then there are the films, books, television documentaries, and radio programs, not to mention the T-shirts and photograph collections on sale.

But there is a very good reason why those involved rarely own up to the deception, and that motive is not necessarily the tax man. The main reason is that, despite open hostility between the crop-circle believers and the circle makers, usually in the direction of the circle makers, the two opposing factions are completely dependent upon each other, because, as any artist will tell you, national publicity is hard to come by.

Without the suggestion of unknown forces at work, very few people would take an interest in crop-circle art on its own merit, so artists need the mythmakers to continue to be as vocal as possible every time a circle appears. Equally, without the circle makers, there would be no circles for otherwise bored individuals to fly over and photograph, or visit en masse with their measuring tapes and electronic devices for gauging unusual electrical activity.

In the debate about whether crop circles are man-made or created by supernatural forces of some kind, the balance of evidence tips very much toward the former. The circle makers have proved pretty conclusively that they are able to create elaborate and complicated designs using relatively simple resources— string, planks of wood, plastic piping, and a garden roller—in a matter of a few hours. They have actually been filmed doing it, but the believers, while accepting that some circles are man-made, still prefer to wonder why their mobile phones fail to work in crop circles, or why flattened wheat is less electrically charged than the standing wheat nearby. Any schoolchild with a basic understanding of physics could step forward and enlighten them by explaining that standing wheat will act like an aerial and attract the atmosphere's electromagnetic charge better than the flattened stems. Presumably that is why all lightning conductors point upward from the roof of your house and do not lie flat in the back garden. And, what's more, my mobile phone doesn't seem to work anywhere in the countryside, let alone in the middle of a field.

All the pseudoscience offered as solid evidence simply doesn't stand up to detailed cross-examination. Nothing has been suggested to prove that crop circles are made by anything other than man himself. There is absolutely no credible evidence of mysterious forces at work, and as is always the case when it comes to proving such things, we will need to see a real alternative to the "man in a cap with a plank of wood working at night" principle. Don't expect many people to believe that ancient ruins under the ground are responsible, or mini tornadoes, plasma vortexes, or any other freak of nature, because if any of these could have created the circles, then it should be easy enough for scientists to prove, or at least reconstruct under controlled conditions. But nobody ever has.

Already I can hear the believers sharpening their tongues in order to dismiss me as a CIA plant or part of a wide-ranging government cover-up program denying the existence of extraterrestrial forces and/or denying the coded messages left in wheat fields by visitors from Mars. That is what the crop believers usually do to vocal opponents of their mystic beliefs, and I am already looking forward to discussing it. Because there is no tangible evidence of any intergalactic interference in our wheat fields, and the only slight piece of evidence ever offered always turns out to be a hoax, later admitted by the hoaxers themselves.

Usually they are the very same circle makers proving to the world how easily fooled the experts are and showing us exactly how they created them. In one such case, from 1996, called the "Oliver's Castle Video," balls of light, referred to by the experts as BOLs, were filmed floating across a field while a crop circle mysteriously appeared directly underneath and the cameraman was heard to whisper, "Wow, that's amazing!"

Never having been in that position myself, I do not know how

I might react if I saw such a thing happen right in front of me, but I imagine it is rather more likely that I would be running down the lane screaming in terror, having dropped all my equipment. Equally suspicious is the way in which the camera stays fixed on the field where the crop circle appears, whereas most cameramen would tend to follow the balls of light with their lens, not hold the camera in one position as the BOLs floated in and out of view.

Further investigation revealed that it was indeed a hoax. John Wheyleigh, a young man from Bath, had created the illusion by filming a wheat field and using an editing program to create the BOLs and then gradually faded out some of the wheat to leave the effect of a crop-circle design. The film caused a sensation and enthusiasts all over the world tried to contact Wheyleigh, but without success.

Digging a little deeper, it came as no surprise to find that "Wheyleigh" wasn't his real name. The young man in question actually turned out to be one John Wabe, a partner in a video-editing company. Needless to say, he sold his video, made a documentary about how he had created his film, and apparently signed a lucrative contract with a television company. Predictably, he has made himself thoroughly unpopular with the more resentment-prone members of the crop-circle and UFO communities across the world, some even threatening to sue him. Others, meanwhile, quietly ignored the hoax and carried on with their important research into intergalactic "messages" left in fields of wheat.

The believers dismiss the evidence of circle makers as the "Doug and Dave effect." Television documentaries about man-made circles are known as "Doug and Dave–style programs," and so on. So blinkered have some of them become that any suggestion of a circle being man-made is derided. (Skeptical believers—now there's a conundrum . . .) In the meantime they have given their "science" a special name. Cereology, they call it, and no, neither I nor the Microsoft spellchecker had heard of that word either.

I am already ordering my insultproof vest in preparation for the publication of this book, because I have a feeling the best reaction I can expect from the community of cereologists is to be called an idiot. We shall see. Of course, as with all of these types of mystery, it is simply impossible to prove a negative. Some people will believe whatever they want to believe, although most of us need to see the firm evidence first, so—as with the Bigfoot mystery (see page 25)—please show us a carcass; some real evidence.

One company, called "Circlemakers" and run by the British artist and documentary filmmaker John Lundberg, makes no attempt to hide its identity, or its work. The company even accepts commercial commissions to create crop circles and has done so all over the world. In one case it re-created a well-known cereal company logo, and in another *The Sun* newspaper asked it to make a huge crop logo of the five Olympic rings to support Britain in its bid to hold the next Games. The following day they ran a front-page headline, "Aliens Back Our Bid," and printed the photograph underneath. Just stop for a minute and wonder how many people in Great Britain, or even across the world, actually believed that headline. Very few, I imagine, but I expect some dyed-in-the-wool cereologists did, even so. When asked why he does it, John has stated that among the numerous reasons for creating crop circles, the chief one is "being able to construct something that most people believe to be beyond human capability."

Now, for me, that is a pretty good reason to do anything, and good luck to him. On their website (www.circlemakers.org) the group claim the circles they create are actually "genuine" in the sense that there is no attempt on their part to deceive anybody.

They are open about their art and ridicule many of the so-called crop-circle "experts" who claim to have had visits from outer space or other paranormal experiences. Well, you would too if you had spent a hard night in a wheat field constructing a giant spiral spelling out the words SHREDDED WHEAT only for somebody to claim it to be the work of little green men from Mars.

So, of the many explanations for the sudden appearance of elaborate designs found in some fields of wheat, ranging from the paranormal to the extraterrestrial, none of them have ever been supported by any genuine evidence. And therefore none of them are as convincing as the most likely explanation—a man in a cap with some string and a plank of wood, plus a flask of tea to keep him going—which is continually dismissed by the cereologists.

So, now I have changed my mind. I began by believing the circle makers were a bloody nuisance and wanted to find out what, if anything, had created the circles not identified as man-made. But instead the only solid evidence I can find is that people have created all of them, so now I respect the circle makers' art, for art's sake, and hope crop circles continue to appear in more and more clever and elaborate forms—and some of them are very clever indeed. The meditation groups down in Sussex who sit in crop circles contemplating whatever it is they contemplate can happily continue to do so as far as I'm concerned, although I think the artist should charge them a fee for it in future. Perhaps he/she could leave out a saucer for the money to be placed in. Although, on second thought, that might create even more confusion.

But for the many who dismiss the circle makers as publicity seekers and hoaxers, I have another idea. Imagine H. G. Wells's time machine, only from outside the machine rather than inside it; in other words, we just happen to be walking past the site of the inventor's house as he flashes through our time zone on his way to the year 30,000 or whenever. You wouldn't see the actual time machine, as it would be traveling too quickly, but its track or footprint would suddenly appear right in front of you, then gradually fade away over the next few weeks. There you are: that's my alternative explanation for crop circles. Scientists of the distant future have managed to build time machines and these are racing backward and forward through our own time zone leaving the footprint of their time machines in our fields, where they are actually standing in thousands of years' time. That would explain why the birds fly around them too. Now, is that any more ridiculous than any of the other theories you have heard from the real experts? And I just made that up, for fun.

But in the meantime the two opposing groups should, in my humble opinion, try to get on with each other. The artists should be allowed to continue creating their art without having their cars vandalized by the believers, and the believers should be allowed to run around in a field measuring bent wheat straws and taking soil samples without people like me making fun of them. And as to that, I really will try to restrain myself in future, but I can't promise anything. (You can call me Doug from now on, or Dave.)

John Dillinger: Whatever Happened to America's Robin Hood?

··

The story of the charismatic criminal who leaped over
counters Hollywood-style when robbing a bank

During the Depression of the 1930s, many Americans, broke
and hungry, made heroes of the outlaws who simply pulled out
their guns and took what they wanted. This was the era of the
gangster: of Al Capone, Bonnie and Clyde, and, most of all, John
Herbert Dillinger.

A career criminal, Dillinger is often described as an American
Robin Hood—although he conveniently skipped the bit about
giving anything back to the poor. Dillinger is best known for his
narrow getaways from police and his many bank robberies
where, incidentally, he also picked up the nickname "Jackrabbit"
because of the athletic way he leaped over counters (supposedly
inspired by something he had seen in a movie).

He was finally cornered by FBI agents at the Biograph Theater in Lincoln Park, Chicago, on July 22, 1934. He had been
there to watch the film *Manhattan Melodrama* with his girlfriend,
Polly Hamilton, and a brothel owner called Anna Sage, who was
facing deportation charges. Sage had cut a deal with the FBI
and, as they exited the theater, she tipped off agent Melvin
Purvis, who gunned Dillinger down from behind.

J. Edgar Hoover, founder and director of the FBI, had become obsessed with capturing the charismatic bank robber, who

was on the run from the Lake County Jail in Crown Point, Indiana, said to be escapeproof. In the quest for the gangster, agents had arrested the wrong man several times and even mistakenly killed three innocent construction workers in a shootout, causing public outrage. Dillinger had been goading Hoover and was becoming something of a Robin Hood–style figure in the eyes of the world. Hoover, in return, was devoting a third of the entire FBI budget to catching this one single outlaw.

But then doubts arose as to whether it was Dillinger who had been shot. It all started when Dillinger's father, summoned to identify the body, failed to recognize his son, famously stating: "That's not my boy." Further investigation appeared to confirm the doubts rather than dispel them. The dead man had brown eyes, for instance, whereas Dillinger's were gray, and the autopsy revealed signs of a childhood illness that he had never had. The corpse also showed signs of a rheumatic heart condition, but Dr. Patrick Weeks, the physician at Crown Point, confirmed

Dillinger had been suffering from no such disease and had been in robust health. Apart from his famed athleticism during bank raids, he had been an avid baseball player both in the navy and while in prison. Furthermore, although fingerprint records were inconclusive because of acid scarring of the hands, the body had none of the scars that had been listed on Dillinger's prison files.

Had the FBI mistakenly killed the wrong man again in their desperate search for John Dillinger? Was he to remain a free man, with J. Edgar Hoover refusing to reveal the truth, as he was already under pressure to resign over the previous mistaken-identity killing? Anna Sage was still deported back to her home country of Romania, leading to speculation she had deliberately misled the FBI by identifying the wrong man, a petty criminal from Wisconsin named Jimmy Lawrence who bore a close resemblance to Dillinger and had dated the same girls. Had John Dillinger found the perfect way to rid himself of Lawrence, a love rival, and the interest of the FBI in one fell swoop? Rumor has it that such was the brazen cheek of the man, he even taunted J. Edgar Hoover by sending him a Christmas card every year afterward.

The Missing Navy Diver

······································

The mysterious disappearance of a real-life
James Bond—the man on whom the
fictional character was based

Lionel "Buster" Crabb, OBE, was the Royal Navy frogman who famously vanished in 1956, when the Suez crisis was at its height, during a reconnaissance mission to investigate a Soviet cruiser.

Crabb's life began uneventfully enough. He was born on January 28, 1909, into a poor family living in Streatham in southwest London. After leaving school he held several menial jobs and then joined the Merchant Navy. At the beginning of the Second World War he joined the army, but it wasn't until he transferred to the Royal Naval Volunteer Reserve in 1941 that he came into his own. In 1942, he was posted to Gibraltar as part of a new Royal Navy diving unit. Their mission was to remove unexploded mines fixed underneath the waterline to the hulls of many Allied ships. It was dangerous, unpleasant work, but Crabb excelled at it. His comrades held his courage and ability in such high regard that they started calling him "Buster" after the American Olympic swimming champion Buster Crabbe (who moved on to a career in the film industry, starring as both Tarzan and Flash Gordon), and the nickname stuck.

His skills were also recognized by his superiors. Buster was awarded the George Medal, promoted to lieutenant commander, and made principal diving officer for northern Italy. At the end of the war he was awarded the OBE for his services to the empire, and posted to Palestine to lead an underwater ex-plosives disposal team removing mines planted by Jewish rebels. In 1947 Crabb left the navy, but he remained in close contact with the military, on one occasion even helping to identify a suit-able location for a nuclear waste discharge pipe for the Atomic Weapons Research Establishment at Aldermaston.

In 1955, as the Cold War gathered pace, the Soviet cruiser *Sverdlov* steamed into Portsmouth harbor as part of a worldwide naval review. Behind the scenes, and the friendly gestures of the world's most powerful nations, Crabb was recruited by naval top brass to make a series of secret dives around the docked *Sverdlov* to evaluate its potential. According to his diving companion Syd-ney Knowles, they found, contained within an opening in the ship's bow, a large propeller that could be directed to give thrust to the bow. Whitehall was impressed, but in the process Crabb had technically become a spy.

In March 1955, Crabb reluctantly retired from professional diving because of his age. The following April, the Russian ship *Ordzhonikidze* arrived in Portsmouth carrying a delegation headed by Soviet leader Nikita Khrushchev. It was during the run-up to the Suez crisis. The British and Egyptian governments were arguing about ownership and rights of access along the Suez Canal; hence, as the Russians were providing the Egyptians with arms, negotiation with the Soviet Union was crucial. So Prime Minister Anthony Eden was both alarmed and dismayed when, without warning, Khrushchev furiously called off the talks, claiming they were being spied upon by British intelli-gence. On his return to Russia, Khrushchev promptly released a statement declaring that his ship's crew had spotted a frogman close to the cruiser berthed in Portsmouth harbor.

Soon afterward the British government issued its own somber statement—that Commander Crabb had been reported missing while "enjoying a recreational dive somewhere along the south coast in Hampshire." This aroused a great deal of suspicion, leading to speculation that perhaps the Russians knew rather more about the baffling disappearance of Britain's best-known diver than the public were being told. And when questions were asked in the House of Commons and Anthony Eden forced Sir John Sinclair, the head of MI6, to resign, it only added to the mystery. After all, if the Russians were this upset over the alleged spying, what information did they have to support it? Could they have captured Crabb?

Compounding the puzzle was the discovery, fifteen months later, of the body of a frogman washed up on a beach at Pilsey Island in West Sussex, just off Chichester. Officials believed it to be that of Buster Crabb but, as the corpse had had both its head and hands cut off, identification was near impossible (using the techniques available at the time). When both Crabb's ex-wife and girlfriend failed to identify the body, there was speculation about yet more shenanigans on the part of the government, brought to a halt when Sydney Knowles was summoned and identified a small scar on the frogman's left knee, thereby confirming that the body was Crabb's.

But the rumors and wild stories continued unabated. In 1961, J. S. Kerans, a British member of Parliament, submitted a proposal to have the case reinvestigated, but this was denied by the Conservative government of the day. In 1964, another MP, Marcus Lipton, made a similar move but with the same result, in the form of a rebuff from a Labour government this time. Some stories suggested that Crabb had been killed by a secret underwater Soviet weapon, while others tried to prove he had been captured and held in Moscow's infamous Lefortovo prison, even citing his prison number (147)—although the Russians strenuously denied this. Another rumor suggested that Crabb had been brainwashed and was now working voluntarily as a specialist instructor for Soviet frogmen. Other accounts maintained he

had deliberately defected to the Russians and was now in charge of the Black Sea Fleet under the name of Lev Lvovich Korablov. Or that he was a secret double agent. Or that—as claimed by Joseph Zwerkin, a former Soviet spy—on being spotted in the water close to the ship, he had been shot by a Russian sniper from the deck of *Ordzhonikidze*.

The strange story of Buster Crabb has intrigued many people over the years. Ian Fleming partly based the character of James Bond on the many colorful tales of Crabb's covert operations. More recently, Tim Binding, author of a fictional account of Crabb's life, *Man Overboard* (2005), claimed he was contacted by Sydney Knowles after its publication. Knowles, then living in southern Spain, apparently met Binding and told him that in 1956 Crabb had intended to defect and that MI5 had become aware of his plans. It would have been a public relations disaster if Commander Crabb—a popular and well-known English war hero, awarded an OBE and with the George Medal pinned to his chest—suddenly became a Soviet citizen. Knowles also alleged that MI5 ordered the *Ordzhonikidze* mission with the sole intention of killing Crabb, even going as far as providing a diving partner to carry out the job. Knowles was then ordered by MI5 to identify the body as Crabb's, despite knowing that the headless corpse wasn't that of his former colleague. The reason he gave for his long silence was that, back in 1989 when he was planning to write an exposé, he had been threatened with death if he continued.

And the confusing events surrounding Crabb's disappearance were only made murkier by the British government's decision to extend the Freedom of Information Act sixty years longer than usual in the case of Buster Crabb. Hence official records will not be made available until the year 2057, one hundred years after the incident. But, based on the evidence that is currently available, this is my interpretation of events.

When the Anglo-Soviet talks were being prepared, Anthony Eden ordered MI5—responsible for overseeing domestic counterintelligence gathering and home security—to do nothing that might cause a diplomatic incident, so crucial was the Suez Canal to British interests. However, this order was not passed to MI6—responsible for overseas security and intelligence. Documents recently released prove that Nicholas Elliott at MI6 recruited Crabb to spy on the *Ordzhonikidze* while it was berthed in Portsmouth harbor. The diver was to gather information about the propeller size and design and check for underwater minelaying hatches. Such information would enable British intelligence to calculate the ship's top speed as well as provide useful information for British torpedo manufacturers.

In April 1956, Buster and his MI6 controller, whose name has been deleted from the records, covertly checked in to the Sally

Port Hotel in Old Portsmouth. On April 19, the pair quietly boarded a small boat and paddled into Portsmouth harbor, where the frogman made a preliminary dive near the Russian ship. He surfaced, briefed the MI6 officer, and then prepared to make a second, more extensive dive. This time, however, Crabb failed to return and was not seen again until his body, minus head and hands—presuming, for the moment, that it *was* his body—was washed up at Chichester. For MI6 to be taking such risks in the first place was an extraordinary development, given that the chances of a British sea battle with the Soviet Union at that time were as unlikely as one with the Portsmouth Yacht Club, and probably about as one-sided too. (Added to which, in the wake of such a diplomatic blunder, Khrushchev delighted in announcing that, far from being a modern state-of-the-art warship, *Ordzhonikidze* was an outdated naval vessel and had been decommissioned. The ship was no longer part of the battle fleet but, instead, was on ceremonial duty ferrying around politicians like him.)

A top-secret memo, now in the public domain, from Rear Admiral John Inglis, director of naval intelligence, denied any official mission by Crabb, stating that if it had been a "bona fide" assignment, there would have been an "immediate and extensive rescue and recovery operation." But, on grounds of diplomatic sensitivity, surely no rescue attempt could seriously have been considered in the waters close to the Russian ship without causing alarm. So was Crabb sacrificed to avoid a diplomatic incident? Or were the Russians already aware of Crabb's presence and did they manage to capture him?

Or did Buster in fact defect, and was the body found at Chichester therefore that of another man? In the wake of the recent defections by middle-ranking diplomats Guy Burgess and Donald Maclean, this would have been a major embarrassment, very much worthy of cover-up by the British government.

It is, after all, known that Nicholas Elliott was responsible for proposing that Crabb should carry out the underwater mission in Portsmouth on that fateful day. Elliott and Kim Philby had been friends at Cambridge when the great twentieth-century spy ring was being formed. But back in 1956, Philby was still seven years away from joining Maclean and Burgess in Moscow. It was not until Elliott confronted Philby in Beirut in 1963, after a defecting Soviet agent had named the latter as a spy, that suspicion arose and connections were made with the Crabb mystery, but somehow Elliott allowed Philby to vanish, only to later reappear in Moscow. It is not officially known what role Elliott played in the spy ring, and some believe that if he was supportive of it, he may well have arranged Crabb's defection in 1956. An alternative theory suggests that Elliott was embarrassed by his connections to Burgess and Maclean, following their defection, and that on learning of Crabb's attempt to do the same, had him murdered by MI6 agents while on his spying mission in the harbor.

Crabb's service to his country appears to have counted for very little in the end, as the government—wishing to avoid a further diplomatic incident—refused to provide his widow with any war compensation, pension, or maintenance payment. Eden's government had been fully aware Lionel "Buster" Crabb was working for the secret service. They lied, both publicly and in private, about events surrounding his disappearance. Many aspects still remain unclear, despite the release of some official documents covering the subject. For example, the official line is that Buster Crabb checked in to the Sally Port Hotel using the name "Mr. Smith." But this couldn't be confirmed, apparently because another secret service agent whose real name was, coincidentally, Mr. Smith and who was also staying at the Sally Port Hotel was enraged to find his name being used and (rather conveniently for the government) tore out the relevant four pages of the hotel guest book.

Assuming Crabb wasn't a turncoat (imagine James Bond defecting!), it would appear that the British government made a mess of their diplomatic relations with the Russians, and in an attempt to whitewash the whole affair, both officially and publicly, made Commander Crabb their scapegoat, washing their hands of him completely. If this is the way Britain treats her war heroes, she doesn't deserve to have any, in my view.

The Dover Demon

....

*What was the bizarre creature spotted on the streets
of Dover, Massachusetts, thirty years ago?*

Late in the evening on April 21, 1977, seventeen-year-old Bill
Bartlett was out driving with two friends along Farm Street in
Dover, Massachusetts. Suddenly something caught the teenager's eye. Slowing to take a closer look, he and the two other
boys assumed it must be some sort of animal crouching by the
wall at the side of the road. Caught in the car's headlights, the
"animal" turned to face them—at which they all froze in terror.
This creature was like nothing they had ever seen before: it had
a small, thin body but a huge oval head and outsize hands and
feet. Trembling with fright, Bill put his foot down hard on the
accelerator and they were off. The boys later described a beast
with large orange eyes and no body hair at all but very rough-
textured skin.

Naturally the authorities were a little skeptical of these crazy
kids, but then another sighting was reported. Two hours after
the first incident, fifteen-year-old John Baxter was walking home
from his girlfriend's when he spotted a strange short figure wandering along the street. He claimed that he then pursued the
creature into a wooded area nearby and observed the beast as it
sat without moving for several minutes. While the cynics among
us would suggest the lad should have stubbed out his illicit cigarette, gone back to bed, and stuck to more plausible excuses in

future, others have pointed out that the description of what he had seen was nearly identical to Bill Bartlett's.

The following evening, Abigail Brabham and Will Taintor reported seeing a "thin and hairless, monkey-like creature" crouching on all fours beside the road. Their description of the creature was the same as that of the others, except in one respect: they insisted that its eyes were green rather than orange, which suggests that the youngsters hadn't collaborated with each other to concoct the whole story. Indeed, they appeared not to know each other. While many similar descriptions have been reported in other parts of the world, there have been no other sightings in Massachusetts.

It is a mystery that has never fully been explained, although, predictably, UFO fans claim the creature was some sort of alien and conspiracy theorists believe it to have been a mutant hybrid, the product of human experimentation that had escaped from a top-secret laboratory. Spiritualists regard it as a being from another realm that accidentally slipped into our own, while zoologists believe the Dover Demon to be nothing more alarming than a baby moose, without explaining the creature's clearly visible hands and feet.

The Dover Demon was only ever seen during that two-day period and no other sightings have ever been reported, which is somewhat surprising, and no evidence has ever been discovered that might identify the mysterious beast.

The Mysterious Disappearance of the Lighthouse Keepers of Eilean Mor

....

What drove three experienced lighthouse keepers to abandon their post on a remote island off the west coast of Scotland?

It was a cold and gloomy afternoon on the Isle of Lewis, and the watchman strained to see the Eilean Mor Lighthouse, located on one of the Flannan Islands, through the mist and rain. Situated on a major shipping route between Britain, Europe, and North America, the rocky Flannans had been responsible for so many shipwrecks over the centuries that the Northern Lighthouse Board had finally decided to build a lighthouse there to warn sailors of the peril.

It had taken four long years to build. But on December 16, 1900, just a year after construction was completed, a report came that the light had gone out. Roderick MacKenzie, a gamekeeper at Uig, had been appointed as lighthouse watchman, and his duty was to alert the authorities if he was unable to see the light. He noted in his logbook that it had not been visible at all between December 8 and 11; he was so concerned, in fact, that he had enlisted the help of all the villagers to take turns watching out for the light, until it was finally seen on the afternoon of December 12.

But when another four days went by and the light failed to appear yet again, MacKenzie alerted the assistant keeper, Joseph

Moore. Moore stood on the seafront at Loch Roag on the Isle of Lewis and stared west into the gloom, looking for the smallest flicker of light, but he too saw nothing. The notion that the brand-new lighthouse might have been destroyed in the recent storms seemed highly unlikely, and at least one of the three resident keepers should have been able to keep the lamp lit, so Moore summoned help.

The following day, owing to high seas, Moore was unable to launch the Board's service boat, the *Hesperus,* to investigate. It would be nine agonizing days before the seas calmed sufficiently for the anxious assistant keeper to leave for Eilean Mor.

Finally, at dawn on Boxing Day, the sky had cleared and the *Hesperus* left Breasclete harbor at first light. As it approached the lighthouse, the boat's skipper, Captain Harvie, signaled their approach with flags and flares, but there was no acknowledgment from the shore. As soon as they had docked at Eilean Mor, the assistant keeper jumped out, together with crew members Lamont and Campbell.

Hammering on the main door and calling to be let in, Moore received no reply. But it was unlocked, so, nervously, Moore made his way inside, to be greeted by complete silence and absolutely no sign of life. The clock in the main room had stopped and everything was in its place, except for one of the kitchen chairs, which lay overturned on the floor.

Moore, terrified of what he might find, was too frightened to venture upstairs until Lamont and Campbell had joined him. But the bedrooms were as neat and tidy as the kitchen and nobody (or indeed no body) was to be seen. The three lighthouse keepers, James Ducat, Donald McArthur, and Thomas Marshall, appeared to have vanished. Ducat and Marshall's oilskin waterproofs were also gone, but McArthur's hung alone in the hallway, in strangely sinister fashion.

Moore saw this as evidence that the two men had gone outside during a storm and that perhaps McArthur, breaking strict rules about leaving the lighthouse unmanned, had raced outside after them. Moore and his fellow crew members then searched every inch of the island but could find no trace of the men. Three experienced lighthouse keepers had seemingly vanished into thin air. Captain Harvie then instructed Moore, Lamont, and Campbell to remain on the island to operate the lighthouse. They were accompanied by one MacDonald, boatswain of the *Hesperus,* who had volunteered to join them.

With that, the *Hesperus* returned to Breasclete, with the lighthouse keepers' Christmas presents and letters from their families still on board, and Harvie telegraphed news to Robert Muirhead, superintendent at the Northern Lighthouse Board: "A dreadful accident has happened at Flannans. The three keepers, Ducat, Marshall, and the Occasional [McArthur in this instance], have disappeared from the Island. . . . The clocks were stopped and other signs indicated that the accident must have happened about a week ago. Poor fellows they must have been blown over the cliffs or drowned trying to rescue a crane [for lifting cargo into and from boats] or something like that." It had been twenty-eight years since the *Mary Celeste* (see page 138) had stirred the public's imagination, and now there was a baffling new mystery to puzzle the world.

In the seventh century A.D., Bishop Flannan, for reasons best known to himself and perhaps his God, had built a small chapel on a bleak island sixteen miles to the west of the Hebrides on the outer limits of the British Isles. The group of islands was known to mariners as the Seven Hunters, and the only inhabitants were the sheep that Hebridean shepherds would ferry over to graze on the lush grass pastures. But the shepherds themselves never stayed overnight on the islands, fearful of the "little men" believed to haunt that remote spot.

The lighthouse on Eilean Mor, the largest and most northerly of the Seven Hunters, was only the second building to be erected on the islands—over a millennium later. Designed and built by David Stevenson, of the great Stevenson engineering dynasty, the building was completed by December 1899, and Superintendent Muirhead of the Northern Lighthouse Board had selected forty-three-year-old James Ducat, a man with more than twenty years' experience of lighthouse keeping, as the principal keeper at Eilean Mor. Thomas Marshall was to be his assistant and the men were to spend the summer of 1899 making preparations to keep the light the following winter.

During that summer, Muirhead joined them for a month and all three men worked hard to secure the early lighting of the station in time for the coming winter. Muirhead later reported how impressed he was by the "manner in which they went about their work." The lighthouse was fully operational for the first time on December 1, 1899. Muirhead's last visit to Eilean Mor before the disappearance was on December 7, 1900. Satisfied that all was well, he then returned to the Isle of Lewis. Although he was not to find out until a few weeks later, the light went out only a day after he had left the island.

When Muirhead returned to join Joseph Moore and the relief keepers on December 29, he brought the principal keeper from Tiumpan Head on Lewis to take charge at Eilean Mor and then began to investigate the disappearance of the three men. The first thing he did was to check the lighthouse journal. He was very perturbed by what he read.

In the log entry for December 12, the last day the lighthouse had appeared to be working, Thomas Marshall had written of severe winds "the like I have never seen before in twenty years." Inspecting the exterior of the lighthouse, he found storm damage to external fittings over one hundred feet above sea level.

The log also noted, somewhat unusually, that James Ducat had been "very quiet" and that Donald McArthur—who had joined the men temporarily as third keeper while William Ross was on leave—was actually crying. And McArthur was no callow youth, but an old soldier, a seasoned mariner with many years' experience and known on the mainland as a tough brawler.

In the afternoon, Marshall had noted in the log: "Storm still raging. Wind steady. Stormbound. Cannot go out. Ship passing sounding foghorn. Could see lights of cabins." This was distinctly odd: no storm had been reported on December 12, and what could possibly have happened to upset an old salt like McArthur?

The following morning, Marshall had noted that the storm was still raging and that, while Ducat continued to be "quiet," McArthur was now praying. The afternoon entry simply stated: "Me, Ducat and McArthur prayed," while on the following day, December 14, there was no entry at all. Finally, on December 15, the day *before* the light was reported for the first time as being not visible, the sea appeared to have been still and the storm to have abated. The final log entry simply stated: "Storm ended, sea calm. God is over all."

Muirhead puzzled over what could have frightened three seasoned veterans of the ocean so greatly, and also what was meant by that last sentence, "God is over all." He had never known any of the men to be God-fearing, let alone to resort to prayer. Equally troubling was where such violent storms had come from when no poor weather, let alone gale-force winds, had been reported in the vicinity at any point up to December 17.

Muirhead also wondered why nobody on Lewis had known of such a frightening storm when the lighthouse was actually visible (bad weather would have obscured it during the day), and for that matter how the passing boat Marshall recorded on December 12 had managed to stay afloat in such a gale. Equally, if it had sunk, why had no boat been reported missing?

Finally, Muirhead wondered if a three-day hurricane raging

over such a localized area was too unrealistic to consider, or simply if one or even all of the lighthouse keepers had gone mad, which might explain the unusual emotions recorded in the lighthouse log and the men's subsequent disappearance. He could think of no other reason for them to disappear on the first calm and quiet day following the alleged storm. If they were going to be swept out to sea, surely that would have been more likely to happen during the gale, had they been foolish enough to venture outside, rather than during the spell of calm weather reported in the final log entry.

One interesting thing to note was that the log that week was written by Thomas Marshall, the second in command and youngest of the three men. That was not so unusual, but for him to be making insubordinate comments about his principal in an official log was certainly out of the ordinary. Especially as the log was bound to be read at some point by the Northern Lighthouse Board and, of course, by James Ducat himself. And to record the aggressive McArthur as "crying" when he would also certainly have read the log himself once the storm had passed seems strangely foolhardy. Yet there it was, in black-and-white, in the official lighthouse log. The whole point of such a record is to note times, dates, wind directions, and the like, not to record human

emotions or activity such as praying. The investigators were baffled by this.

Clearly the men on the island had been affected by a powerful external force of some kind, and so Superintendent Muirhead turned his attention to the light itself, which he found
clean and ready for use. The oil fountains and canteens were full
and the wicks trimmed, but Muirhead knew the light had not
been lit at midnight on December 15 because the steamship
Archtor had passed close to the Flannan Islands at that time and
the captain had reported he had not seen the light, when he felt
sure it should have been clearly visible from his position.

The kitchen was clean and the pots and pans had been
washed, so Muirhead concluded that whatever had happened to
the men had taken place between lunchtime and nightfall, before the light was due to be lit. But there had been no storm on
that day, as evidence from both the lighthouse log and the Isle of
Lewis confirms.

Muirhead then decided to make a thorough search of the site
and, despite high seas, was able to reach the crane platform seventy feet above sea level. The previous year a crane had been
washed away in a heavy storm, so the superintendent knew this
to be a vulnerable spot, but the crane was secure, as were the barrels and the canvas cover protecting the crane.

But curiously, forty feet higher than the crane, 110 feet above
sea level, a strong wooden box usually secured into a crevice in
the rocks and containing rope and crane handles was found to
be missing. The rope had fallen below and lay strewn around the
crane legs, and the solid iron railings around the crane were
found to be "displaced and twisted," suggesting a force of
terrifying strength. A life buoy fixed to the railings was missing
but the rope fastening it appeared untouched, and a large, approximately one-ton section of rock had broken away from the
cliff, evidently dislodged by whatever it was that had caused the

rest of the damage, and now lay on the concrete path leading up to the lighthouse.

Muirhead considered whether the men could have been blown off the island by the high winds but decided this would have been impossible during the calm weather of December 15. Further inspection revealed turf from the top of a two-hundred-foot cliff had been ripped away, and seaweed was discovered, the like of which no one could identify. Muirhead thought that a mammoth roller wave could have swept away the two men in oil-skins working on the crane platform, but such a freak wave had never been reported before.

Unable to come to a definite conclusion, Muirhead returned to Lewis, leaving a very uneasy Joseph Moore with the new principal keeper, John Milne, and his assistant, Donald Jack. In the report he made on January 8, 1901, a sad and baffled Muirhead noted that he had known the missing men intimately and held them in the highest regard. He wrote that "the Board has lost two of its most efficient Keepers and a competent Occasional." And he concluded his report by recalling: "I visited them as lately as 7th December and have the melancholy recollection that I was the last person to shake hands with them and bid them adieu."

At the subsequent Northern Lighthouse Board inquiry, also conducted by Robert Muirhead, it was noted that the severity of the storm damage found on Eilean Mor was "difficult to believe unless actually seen." The inquiry concluded:

From evidence which I was able to procure I was satisfied that the men had been on duty up until dinner time on Saturday the 15th December, that they had gone down to secure a box in which the mooring ropes, landing ropes etc. were kept, and which was secured in a crevice in the rock about 110 foot above sea level, and that an extra large sea had rushed up the face of the rock, had gone above them, and coming down with immense force, had swept them completely away.

But this pathetic attempt by the Board fails to explain why McArthur was there without his oilskins and does not account for his disappearance, unless the Board believed he had run to the cliff top and, on finding his colleagues in the sea, thrown himself in after them wearing just his smoking jacket and carpet slippers. The inquiry also makes no reference to the fact that the damage to the railings and landing platform could have been caused *after* the men had gone missing on the fifteenth, possibly even during the heavy storms and gales recorded on December 20.

Later, it came to light that a further piece of evidence had been submitted to the inquiry but was not made public. Two sailors who were passing Eilean Mor on the evening of December 15 claim to have been discussing the lighthouse, and why it was in complete darkness, when they noticed a small boat being rowed frantically across the sea by three men dressed in heavy-weather clothing. By the light of the moon, they watched as the small boat passed close to them and they called out to the men. Their calls were ignored, however, and the boat made its way past them and out of sight.

Over the years, all the usual theories have been trotted out—yes, including sea monsters and abduction by aliens, not to mention the curse of the "little men"—but staying within the realms of reality and on the basis of observations made at the time, only two explanations seem feasible.

The first is that the west landing at Eilean Mor is located in a narrow gully in the rock that terminates in a cave. During high seas or storms, water forced into the cave under pressure will return with explosive force, and it is possible that McArthur, noticing heavy seas approaching, rushed out to warn his two colleagues working on the crane platform, only to become caught in the tragedy himself. This would explain the over-turned chair and the reason he was not wearing his oilskins. Even so, it seems somewhat unlikely that while in such a tearing hurry, McArthur would have paused on his way out to carefully close both of the doors and the gate to the compound.

The second theory is that one man in oilskins fell into the water and the other rushed back to the lighthouse to call for help. Both men then fell in while attempting to rescue the first. But once again this theory fails to explain the closed doors and gate, and is not consistent with the sighting of three men in a boat by moonlight. In 1912, a popular ballad called "Flannan Isle" by William Wilson Gibson added to the mystery by offering all sorts of fictional extras, such as a half-eaten meal abandoned in a hurry—conjuring up images of the *Mary Celeste*. But this only clouds the very real tragedy of three men losing their lives on a bleak, windy rock in the North Sea while working to pre-vent others from losing theirs.

Following the terrible and mystifying events, the lighthouse nonetheless remained manned, although without incident, by a succession of keepers, and in 1925 the first wireless communica-tion was established between Eilean Mor and Lewis. In 1971, it was fully automated, the keepers withdrawn, and a concrete helipad installed so that engineers could visit the island via less hazardous means for annual maintenance of the light. Nobody has lived on Eilean Mor since.

The most plausible theory arose by accident nearly fifty years after the disappearance of the lighthouse keepers. In 1947, a Scottish journalist named Iain Campbell visited the islands and, while standing on a calm day by the west jetty, he observed the sea suddenly heave and swell, rising to a level of seventy feet above the landing. After about a minute the sea returned to its normal level. Campbell could not see any reason for the sudden change. He theorized it may have been an underwater seaquake

(see also "Whatever Happened to the Crew of the *Mary Celeste*?"
page 138) and felt certain nobody standing on the jetty could
have survived. The lighthouse keeper at the time told him that
the change of level happened periodically and several men had
almost been pulled into the sea but managed to escape.

Although this seems the most likely fate of the men on December 15, 1900, it is by no means certain and still fails to explain several known clues, such as why the third man
disappeared wearing his indoor clothing after carefully closing
and latching three doors behind him, or who the three men in
the rowing boat could have been. Nor does it account for the
strange logbook entries or why the light appeared not to be operational for a number of days. The only thing we know for certain is that something snatched those three brave men off the
rock on that winter's day over a hundred years ago, and nothing
has been seen or heard of them since.

Fairies at the Bottom of the Garden

....

Do you believe in fairies—or in photos of them?
The intriguing pictures taken by two young
girls in Cottingley, Yorkshire.

One of the most famous mysteries of the twentieth century is the story of the Cottingley Fairies, photographed by sixteen-year-old Elsie Wright and her ten-year-old cousin Frances Griffiths during the summer of 1917. The series of pictures were taken over a period of two months at a small brook near their home at Cottingley, a picturesque village in Yorkshire.

Young Frances swore that she had seen the fairies and her elder cousin confirmed her story. The photographs, they said, proved their outlandish claims. Harold Snelling, an expert in fake photography, declared that "these dancing figures are not made of paper nor any fabric; they are not painted on a photographic background—but what gets me most is that all these figures have moved during the exposure." He seems not to have spent much time considering the possibility that wind, either natural or created, might have moved the images. Sir Arthur Conan Doyle, by then world famous for having created Sherlock Holmes, lent considerable credibility to the story when he stated that he believed the photographs were "genuine beyond doubt," and he even wrote a book, *The Coming of the Fairies*, which convinced many people that the tiny figures in the photos were real. (For more on Conan Doyle, see "The Spine-chilling Tale of the

Chase Vault," page 39, and "Whatever Happened to the Crew of the *Mary Celeste*?" page 138.)

It wasn't until many years later, not until 1983 in fact, that the elder of the two girls admitted that the photographs had been faked and made with pictures cut out of a glossy magazine and held together with pins. On the other hand, Frances always maintained she had seen the fairies with her own eyes and swore that the fifth picture, showing fairies in a sunbath, was 100 percent genuine.

The Mystery of Our Lady of Fatima

*The visions of the Virgin Mary seen by three
Portuguese peasant children and the extraordinary
"Miracle of the Sun" witnessed by thousands*

On July 13, 1917, three children were startled to find a mysterious figure approaching them as they tended their flock of sheep in pastureland near Fatima in Portugal. Lucia Rosa dos Santos and her two cousins Jacinta and Francisco Marto reported seeing what they described as a "pretty lady from Heaven." Lucia said the lady was "brighter than the sun, shedding rays of light clearer and stronger than a crystal glass filled with the most sparkling water and pierced by the burning rays of the sun." Just the sort of description you'd expect from an illiterate ten-year-old shepherd girl. Lucia also claimed the lady had entrusted her with three important secrets, which she did not reveal until many years later.

Instead of being cuffed around the ear, the three scallywags were firmly believed, and the devout soon identified the mysterious visitor as the Blessed Virgin Mary herself. Word of the vision rapidly spread, and thousands began making the pilgrimage to the Cova da Iria (the area of pastureland near Fatima in which the children had grazed their sheep) hoping to see the Mother of Jesus for themselves. Artur de Oliveira Santos, the mayor of Vila Nova de Ourem and the most powerful man in the region, became increasingly anxious about the political implications of the pilgrimage to Fatima. Reports of new miracles were only swelling the number of pilgrims. His open hostility to the alleged apparitions was well known, and he ordered the arrest of the little ones. On August 13, the children were arrested on their way to the pasture at Cova da Iria and thrown into jail. Other prisoners later testified the youngsters were initially frightened and upset, but were soon chanting their rosaries and leading their cellmates in prayer.

When Santos interrogated the children, they wouldn't tell him anything. So, as the story goes, he arranged for a large pot of boiling oil to be delivered to the interrogation room. He then took the children one by one to the room, claiming that each of the others had been boiled to death in oil for "failing to tell him the truth." The "remaining" child was urged to speak out or suffer the same fate. Remarkably, despite such persuasive techniques, the psychopath still failed to persuade the youngsters to tell him anything at all.

With that, Santos was forced to release them. Six days later, on August 19, they reported another visitation at nearby Valinhos. On September 13, the Blessed Virgin appeared in the field again, and this time the children reported she promised them that at noon on October 13 she would reappear and perform a miracle, so that "everybody will believe."

As dawn broke on October 13, a thick layer of cloud hung over the entire area and heavy rain fell, soaking the thousands who had gathered to see the expected miracle. Many were present only to witness what they were sure would be a nonevent. The tension mounted as crowds of between seventy and one hundred thousand gathered during the morning. People from every walk of life were there, including doctors, lawyers, and scientists (not normally inclined to be credulous), religious leaders and the great and the good, all eagerly awaiting the great event. Noon passed without incident, but in the middle of the afternoon tens of thousands of people witnessed the clouds gradually part to reveal a dim, opaque sun spinning on its axis and emitting various bright colors that illuminated everything around. After a short while the sun apparently began to detach itself from the sky and plummet toward the earth, but instead of crashing to the ground and wiping out the entire human race, it slowed down at the last moment and only came close enough to heat the land and dry out everybody's soaked clothing before slowly making its way back to its regular place in the sky.

This event, which lasted for between eight and ten minutes and in which the sun appeared to sink and rise again three times, became known as the "Miracle of the Sun." Previously a strongly Catholic country, Portugal at that time had been a secular state for only seven years—since the monarchy had been abolished during the republican revolution of 1910. Since then, the new government of Portugal had been severely hostile toward religious groups, which explains Mayor Santos's unpleasantness to the devout children. However, even the pro-government *O Seculo*, Portugal's most influential newspaper, was unable to repress its excitement on this occasion. Popular columnist Avelino de Almeida noted:

> . . . the astonished eyes of these people, whose attitude takes us back to biblical times and who were white-faced with shock and with their heads uncovered, facing the blue sky. The sun has trembled, the sun has made sudden movements that were outside all cosmic laws—the sun has "danced," according to the typical expression of the country people. Covered with dust on the running board of the bus from Torres Novas, an old man recites the Creed, from beginning to end. I ask who it is and they tell me it is João Maria Amado de Melo.
>
> I see him later talking to those around him, who still have their hats on, begging them strongly to take them off in the presence of such an extraordinary demonstration of the existence of God. Identical scenes are repeated in other places

and a woman shouts, bathed in tears and almost suffocated, "What a shame! There are still men who don't take off their hats in the presence of such a miracle."

And next they ask each other if they have seen it or not. Most confess that they have seen the dancing of the sun but others, however, declare they have seen the smiling face of the Virgin herself. They swear that the sun spun about itself like a ring of fireworks and that it came down almost to the point of burning the Earth with its rays. Some say that they saw it change color. It was about three in the afternoon.

De Almeida claimed to have witnessed the whole event, but the photographer standing next to him, Judah Ruah, nephew of the famous shutterbug Joshua Benoliel, said he saw nothing at all. When asked why, he replied, "because nothing strange happened to the sun. But when I saw all those people kneeling I understood something to be happening and so I photographed them instead."

Another journalist, from the Lisbon newspaper *O Dia,* reported:

The silver sun, enveloped in the same gauzy gray light, was seen to whirl and turn in the circle of broken clouds. The light turned a beautiful blue, as if it had come through the stained-glass windows of a cathedral, and spread itself over the people who knelt with outstretched hands. People wept and prayed with uncovered heads in the presence of the miracle they had awaited. The seconds seemed like hours, so vivid were they.

An eminent eye surgeon, Dr. Domingos Pinto Coelho, recorded that "the sun, in one moment was surrounded with scarlet flame and at another aureoled in yellow and deep purple. It seemed to be in an exceedingly fast and whirling movement, at times appearing to be loosened from the sky and to be approaching the earth and strongly radiating heat." There can be little doubt *his* eyes were not deceiving him (or they shouldn't have been at any rate). Another medical man, Dr. Almeida Garrett of Coimbra, wrote that "the sun, whirling wildly, seemed to loosen itself from the firmament and advance threateningly upon the earth as if to crush us with its huge and fiery weight. The sensation was terrible." And another learned individual, Dr. Formigao, a professor at the seminary at Santorem, noted that "suddenly the rain stopped. The clouds were wrenched apart and the sun appeared in all its splendor. Then it began to revolve on its axis like the most magnificent fire wheel that we could imagine, taking all the colors of the rainbow and sending forth multicolored flashes of light, producing the most astounding effect."

With an estimated one hundred thousand people present, the weight of witness evidence is overwhelming. This, coupled with the children's ability to predict the event to within a few hours, proved to many people that they truly had experienced a miracle. Even so, a careful examination of individual statements reveals many contradictions. In some the sun looked like a "ball of snow" and in others an opaque disc. Some reports state the sun was "dancing" and in others that it was zigzagging. Some witnesses believed it actually touched the earth's surface, while others failed to see it move at all. There have been statements claiming columns of fine blue smoke, and others describing how the very air seemed to change color. Everybody claims to have witnessed the miracle at the same time and simultaneously let out either a roar or loud gasp that echoed around the valley, but the timing of the reports varies from between midday and dusk.

The only connecting theme is that most people saw something happen at around the same time and on the same afternoon. But scientific records contain no reports from anywhere

in the world of unusual astronomical or solar activity. This is strange, because even if there had been a natural reason for the phenomenon, such as a cloud of dust from the Sahara Desert or unusual atmospheric gases, as has been suggested, then astronomers would have recorded these. That is exactly what happened when stratospheric dust made the sun appear to be blue and red to the people of China in 1983, or when the blue moon was seen for two years after the eruption of Krakatoa in 1883.

It is hard to believe that so many well-educated and rational people could either make their story up or be fooled by what might amount to a vast con trick. But for those who claim some sort of collective hallucination at Fatima, there is the evidence of witness reports from as far away as thirty miles from people going about their normal business.

Some years later, in 1931, Lucia claimed Jesus himself had visited her in Rianxo, Galicia, to teach her two new prayers and he had given her a message for the pope. Soon afterward the Catholic Church added its considerable weight to the debate by announcing they were "approving the visions as worthy of belief." So, with the official stamp of approval from the Vatican, the

Blessed Virgin as she appeared at Cova da Iria became known throughout the world as Our Lady of Fatima. By then little Lucia de Jesus Rosa dos Santos had become Sister Lucia of Jesus, following her ordination as a Carmelite nun.

In 1942, as the Second World War was at its bloodiest, Sister Lucia finally decided to reveal the first of the secrets confided in her by the Virgin Mary all those years ago. The first was a terrifying vision of Hell, which she recorded in her third memoir, published toward the end of that year:

> Our Lady showed us a great sea of fire which seemed to be beneath the earth. Plunged in this fire were demons and souls in human form, like transparent burning embers, all blackened or burnished bronze, floating about in the conflagration, now raised into the air by the flames that issued from within themselves together with great clouds of smoke, now falling back on every side like sparks in a huge fire, without weight or equilibrium, and all the time the shrieks and groans of pain and despair, which horrified us and made us tremble with fear. The demons could be distinguished by their terrifying and repulsive likeness to frightful and unknown animals, all black and transparent.

Fortunately, as she goes on, "This vision lasted but an instant. How can we ever be grateful enough to our kind heavenly Mother, who had already prepared us by promising, during the first Apparition, to take us to heaven? Otherwise, I think we would have died of fear and terror."

The second secret included Mary's instructions on how to save souls from Hell and convert the world to Christianity:

> You have seen hell where the souls of poor sinners go. To save them, God wishes to establish in the world devotion to my Immaculate Heart. If what I say to you is done, many souls will be

saved and there will be peace in the world. The war is going to end, but if people do not cease offending God, a worse one will break out during the Pontificate of Pius XI. When you see a night illumined by an unknown light, know that this is the great sign given to you by God that he is about to punish the world for its crimes by means of war, famine, and persecutions of the Church and of the Holy Father. To prevent this, I shall come to ask for the consecration of Russia to my Immaculate Heart and the Communion of Reparation on the First Saturdays. If my requests are heeded, Russia will be converted and there will be peace. If not, she will spread her errors throughout the world, causing wars and persecutions of the Church. The good will be martyred; the Holy Father will have much to suffer and various nations will be annihilated. In the end, my Immaculate Heart will triumph. The Holy Father will consecrate Russia to me, and she shall be converted and a period of peace will be granted to the world.

All of you who find the timing and content of this revelation suspicious, what with it coming in the middle of a "worse" war and during Pope Pius's reign (even if it was Pius XII rather than XI by this time), will be struck down by the next bolt of lightning. But come on! This is a serious matter. The Catholic Church actually approved the visions as "worthy of belief," and granted them genuine, bona fide miracle status, eleven years before revealing one of the three big "secrets" with its anti-Russian warning. Are they really expecting us to believe the Virgin Mary visited three young peasant children in Portugal in 1917 to warn the world about something as specific and politically biased as the threat from Russia in twenty-five years' time?

For reasons best known to themselves, Vatican officials refused to release the third secret until the late 1990s, when Pope John Paul II finally unveiled Sister Lucia's (somewhat disjointed) account:

After the two parts which I have already explained, at the left of Our Lady and a little above, we saw an angel with a flaming sword in his left hand. Flashing, it gave out flames that looked as though they would set the world on fire; but they died out in contact with the splendor that Our Lady radiated toward him from her right hand: pointing to the earth with his right hand, the Angel cried out in a loud voice: "Penance, Penance, Penance!" And we saw in an immense light that is God: "something similar to how people appear in a mirror when they pass in front of it," a Bishop dressed in White, "we had the impression that it was the Holy Father." Other Bishops, Priests, men and women Religious [*sic*] going up a steep mountain, at the top of which there was a big Cross of rough-hewn trunks as of a cork-tree with the bark [*sic*]; reaching there the Holy Father passed through a big city half in ruins and half trembling with halting step, afflicted with pain and sorrow, he prayed for the souls of the corpses he met on his way; having reached the top of the mountain, on his knees at the foot of the big Cross he was killed by a group of soldiers who fire [*sic*] bullets and arrows at him, and in the same way there died, one after another, the other Bishops, Priests, men and women Religious, and various lay people of different ranks and positions. Beneath the two arms of the Cross there were two Angels each with a crystal aspersorium [a vessel for holding holy water] in his hand, in which they gathered up the blood of the Martyrs and with it sprinkled the souls that were making their way to God.

It is said that Pope John Paul II believed that the text refers to the attempt on his life in St. Peter's Square by Mehmet Ali Agca in 1981, while others have suggested it is a prediction of the end of the world. I, to be honest, can make neither head nor tail of it.

In 1946, Sister Lucia joined the Convent of the Carmelite Sisters at Coimbra, where she remained until her death on February 13, 2005, passing away shortly before her ninety-eighth birthday. During her lifetime she wrote six memoirs and two books. The day of her funeral, February 15, was declared a national day of mourning, even disrupting political campaigning for the Portuguese parliamentary elections a few days later. Before she died, Sister Lucia claimed to have seen her "pretty lady from Heaven" many times throughout her lifetime, although nobody else ever witnessed or corroborated her claims. I wish I could. In fact I only wish the Blessed Virgin would appear round here. We could do with a few heavenly visitations to pep things up a bit.

The events in Fatima during 1917 have never been fully explained and remain as mysterious today as they did in the beginning. Nobody knows what happened, although everybody agrees that something did. I find it hard to completely rule out the appearance of some sort of vision or natural phenomenon occurring near Fatima on those four separate occasions because that would suggest that around one hundred thousand people were either mentally ill, deluded, or simply lying. The sightings were, however, made in the midst of the Great War, when things looked particularly gloomy. Perhaps people so badly needed evidence of something spiritual, some proof of divine interest in them, that they were able to convince themselves that they had witnessed something rather more impressive than they actually had. After all, the famous incident of the Angels of Mons (the supposedly supernatural force of ghostly warriors that intervened to help protect British forces at the crucial moment in the battle of Mons), which had happened only a couple of years earlier, is commonly considered now to be a mixture of morale-boosting propaganda and hallucinations on the part of sleep-deprived soldiers. And we have seen how it is possible to create an illusion on a huge scale, as evidenced by the way in which magician David Copperfield made the Statue of Liberty appear to vanish before the eyes of millions of people, and yet we know it didn't really go anywhere.

So that leaves us with one final question. Even without the assistance of "magic," is such mass deception otherwise possible—convincing a vast group of people to believe in the same lie at roughly the same time? Anybody would have to hesitate before saying yes. But then you think of Tony Blair's second and third election victories and you realize that of course it's possible; in fact it's surprisingly easy. Or to quote Abraham Lincoln, it's perfectly feasible to "fool all of the people some of the time."

What Happened to the Lost King of France?

....

Did Louis-Charles, heir to the French throne,
really survive or was someone just pretending?

After the French royal family had been cut down to size—by the guillotine-wielding revolutionaries in 1793—a story about Louis-Charles, eldest son of Louis XVI and Marie Antoinette and heir to the throne, quickly spread throughout Europe.

According to the official version of events, the eight-year-old had been separated from his parents and elder sister at the Temple prison in Paris and had been incarcerated on his own in an attempt to prevent loyalists from rescuing the boy and reestablishing the monarchy. To make the point that he was now just one of the people, his captors called him "Louis Capet"—after his ancestor Hugh Capet, founder of the royal dynasty, but also as a deliberate insult, as royalty tend not to use surnames—and set him to work as a cobbler's assistant. The former dauphin was also forced to sing revolutionary songs, drink alcohol, and curse his mother and father. He remained in prison for two years, dying of tuberculosis in 1795.

But when his death was announced, a story circulated the courts of Europe that soldiers loyal to the king had substituted a dying peasant boy for the royal lad and he had been spirited away to safety and to await his coming of age and a suitable moment to retake the throne. It was suggested that the dauphin might have been smuggled out in a bathtub: a guard claimed

that one of the men carrying a tub of water from the dauphin's room stumbled and the cry of a young boy could clearly be heard.

The doctor who performed the autopsy on the dead boy removed his heart, as was common at the time when a member of the royal family had died, and pickled it in alcohol—presumably to keep the royal livers company on the shelf. Ten years later, one of his students stole the jar and kept it hidden until his own death, when his wife sent it to the archbishop of Paris.

In 1814—shortly before Napoleon, escaping from exile in 1815, was thoroughly routed by the Duke of Wellington (not relevant at all to the story, but I like to remind people anyway)—the Bourbon monarchy was restored in the person of Louis XVIII, brother of Louis-Charles, or Louis XVII as he would have been. At the time, hundreds of claimants to the throne, all professing to be the "lost dauphin," arrived in Paris from all over Europe, some from as far away as Canada, South Africa, and the Seychelles.

Of these only one seemed plausible to many royalists, a German clockmaker by the name of Charles-Guillaume Naundorff, who had mysteriously appeared in Berlin during 1810, seemingly from nowhere. His claim was supported by proof that his age matched the birth date of the real dauphin, but very little else could be established as he had no birth certificate and no proof of who his parents were. Some claimed him to be the son of Marie Antoinette and her lover Axel de Fersen, while others dismissed him as an impostor, and he never managed to establish a true claim to the French throne. Nonetheless, his death certificate, issued in Holland, named him as Louis-Charles of Bourbon, Duke of Normandy, the correct form of address known only in royal circles. His tomb in Holland bears the inscription: "Here lies Louis XVII, Charles Louis, Duke of Normandy, King of France and of Navarre." Subsequent forensic and DNA testing of his remains have proved inconclusive, however.

The pickled heart many believe to be that of Louis XVII passed through several hands between 1830 and 1975, when it was finally laid to rest in the royal crypt in the Saint Denis Basilica, close to the remains of Louis XVI and Marie Antoinette. Even then, there was a challenge to its authenticity by one of the descendants of Naundorff, his great-great-grandson Charles Louis Edmond de Bourbon, who fought to assert his claim to the title of prince. To this day nobody has satisfactorily confirmed whether Naundorff was in fact a prince or a prat, or if there is much difference anyway.

The Strange Case of Kaspar Hauser

*Who was the foundling boy discovered in
Nuremberg, Germany, and was he really
as backward as he first appeared?*

On the morning of May 26, 1828, a boy approximately sixteen years of age and dressed in rags was found wandering the streets of Nuremberg. He appeared unable to speak except to say "horse" and "I want to be a soldier like my father." (Although that is about as much German as I would like to know myself, it didn't help him explain his unusual circumstances.) In the street he approached a shoemaker called Weissman and handed him the two pieces of paper he carried in his pocket. The first was dated October 1812 and appeared to be a letter from his mother:

> This child has been baptized and his name is Kaspar. You must give him his second name yourself. I ask you to take care of him. His father was a cavalry soldier. When he is seventeen, take him to Nuremberg, to the 6th Cavalry Regiment; his father belonged to it. I beg you to keep him until he is seventeen. He was born on 30th April 1812. I am a poor girl; I can't take care of him. His father is dead.

The second letter was undated. It read:

Honored Captain,

   I send you a lad who wishes to serve his king in the Army. He was brought to me on 7th October 1812. I am but a poor laborer with children of my own to rear. His mother asked me to bring up the boy, and so I thought I would rear him as my own son. Since then, I have never let him go one step outside the house, so no one knows where he was reared. He himself does not know the name of the place or where it is.

You may question him, Honored Captain, but he will not be able to tell you where I live. I brought him out at night. He cannot find his way back. He has not a penny, for I have nothing myself. If you do not keep him, you must strike him dead or hang him.

The shoemaker took Kaspar to the town magistrate, who passed him into the care of Andreas Hiltel, a jailer at the Vestner Gate Tower, who placed him in his own private living quarters. To Kaspar's obvious delight, many curious people came to visit him. They found a boy appearing to have a mental age of about six, who walked barely better than a toddler and could eat only bread and water. Hiltel observed the lad closely and noticed that despite these deficiencies, he had an excellent memory, which led Hiltel to think Kaspar was of noble origin. (It's never seemed to me that the posher you are the better you are at remembering things, but maybe things were different back then.) Over a period of three months Hiltel patiently taught Kaspar enough words for the boy to communicate what had happened to him.

It appeared that Kaspar, for as long as he could remember, had been kept in a dark two-meter-square cell with nothing but a straw bed and a wooden horse for a toy. Bread and water were placed in the room through a small hatch. Sometimes the water tasted bitter—suggesting it had been drugged—and on these occasions he fell asleep and woke to find his hair and nails had been trimmed and his clothes changed. The first human contact of his life came when a man opened the door of his cell and led him outside. Kaspar said he then fainted and woke up to find himself on the streets of Nuremberg.

Following his time at Vestner Gate, Kaspar was given into the care of schoolteacher Friedrich Daumer, who took a close interest in him and taught him to speak, read, and write. Over the ensuing year, the lad developed into an intelligent and likable young man who appeared to thrive in his new environment.

Then, on October 17, 1829, a hooded man attacked Kaspar in the street and tried to stab him, succeeding only in wounding him in the forehead. Officials quickly moved Kaspar into the care of Baron von Tucher, who found him employment at a local law office. But why had someone tried to kill him?

Because of a faint family resemblance, rumors had begun to circulate that Kaspar was in fact the son and heir of Karl, Grand Duke of Baden, and Stéphanie de Beauharnais, daughter of Napoleon Bonaparte. According to these rumors, Duchess Stéphanie had given birth at around the time Kaspar would have been born but the baby was quickly taken from her bedchamber. She was later told that her child had died. Because the duke then appeared to have no heir, his successor was to be Leopold I of Baden. This had all been engineered by his mother, the Countess von Hochberg, who had arranged the kidnapping of the duchess's child and the subsequent attempt on Kaspar's life.

Four years later, on December 14, 1833, having been moved to Ansbach, Kaspar was contacted and told he could learn about his ancestry if he went to the Court Gardens. On his way there, Kaspar was attacked again and stabbed in the chest. He survived long enough to stagger home but died only a few days later without being able to identify his assailant.

Kaspar was buried in a small, peaceful graveyard, and his headstone reads: "Here lies Kaspar Hauser, riddle of his time. His birth was unknown, his death mysterious." A monument was later erected to him in Ansbach, which reads, *"Hic occultus occulto occisus est"* ("Here an unknown was killed by an unknown"). His death gave birth to one of Europe's best-known and most enduring mysteries, one that will probably never be fully solved. (See "The Piano Man," page 197, for the story of a modern-day Kaspar Hauser.)

The Great Loch Ness Con Trick

●●●●●●●●●●●●●●●●●●●●●●●●●●●●●●●●●●●●●●●

If the Loch Ness Monster doesn't exist,
how come there have been so many pictures
and sightings? And is Nessie really Nellie?

The first documented sighting of a monster inhabiting Loch
Ness was by St. Columba in A.D. 565. According to this account,
the Christian missionary was traveling through the Highlands
when he came across a group of Picts holding a funeral by the
loch. They explained that they were burying a fellow tribesman
who had been out on the loch in his boat when he had been attacked by a monster. Columba immediately ordered young
Lugne Mocumin, one of his own followers, to swim across the
loch to retrieve the dead man's boat. Detecting lunch was on its
way again, the great beast reared up out of the water, at which
Columba held up his cross and roared: "Thou shalt go no further, nor touch the man; go back with all speed!" And with that,
the terrified monster apparently turned tail and "fled more
quickly than if it had been pulled back with ropes, though it had
just got so near to Lugne, as he swam, that there was not more
than the length of a spear-staff between the man and the beast."
The group of Picts, very impressed by all this, converted to Christianity on the spot. However, as evidence of a monster living in
the loch for the last fifteen hundred years, this account seems
about as reliable as the story of the tooth fairy. Not least because
St. Columba also claimed, a tad implausibly, to have had various
other successful run-ins with Scottish monsters, once even slay-

ing a wild boar just with his voice. Nevertheless, many were convinced by the Loch Ness tale.

Then there was silence on the monster front until some
strange sightings were reported in the eighteenth and nineteenth centuries. But the Loch Ness Monster, as we have come to
know and love it, wasn't really "born" until much later—not
until 1933, in fact, when (prosaically enough) the A82 trunk
road had finally been completed along the western shore of
Loch Ness, connecting the western town of Fort William with
the busy port of Inverness on the North Sea. Providing easy access for tourists and industry alike, the road also offered a route
past the picturesque loch for the first time.

Nearby Inverness had a long-standing and hugely popular tra-
dition of hosting an annual circus. In 1933, Bertram Mills took
his circus to Inverness along the new A82 for the first time,
where his road crew would have stopped along the banks of
Loch Ness to rest and feed the animals. Coincidentally that was
when the sightings of the Loch Ness Monster began. Bertram
Mills, ever the entrepreneur, quickly used the local story to his
advantage by offering £20,000 (nearly £2 million today) to anybody who could prove that they had seen the great beast. It was
a sum Mills seemed suspiciously unable to afford to pay out. But
the public flocked to the area nevertheless, sightings soared,
and more people than ever before attended his shows in case
the monster might make an appearance.

But how could Mills have been so sure nobody could legiti-
mately claim the reward? My theory is that he must have seen the
famous photo of a plesiosaur-like creature taken in 1933 near
Invermoriston by a Scottish surgeon and had known that it was
no monster. At the time, skeptics claimed the photograph was a
fake: the creature it showed was thought to have been an otter
or maybe vegetation floating on the surface of the loch. It was
even said to be an elaborate hoax created using a toy submarine.

But Bertram Mills had seen an elephant swim before and must have realized the photograph taken was most likely of one of his animals bathing in the loch. The financial benefits of staying silent about this were obvious, however.

Soon afterward, on April 14, 1933, a Mr. and Mrs. Mackay claimed that they had seen a "large . . . whale-like beast" idling in the loch and that it had then dived under, causing "a great disturbance" in the water. They had immediately reported the sighting to a local gamekeeper, Alex Campbell. Campbell, conveniently enough, also turned out to be an amateur reporter for *The Inverness Courier.* His embellished account of the sighting, entitled "Strange Spectacle on Loch Ness," appeared on May 2, 1933, and brought him instant fame. The world's monster hunters, not to mention the media, then descended on this remote area of the Scottish Highlands, previously known only for its fishing.

The dial of Loch Ness Monster excitement was then cranked up even further by the *Daily Mail,* when they sent in a professional team of monster hunters headed by the wonderfully named big-game hunter Marmaduke Wetherell. The *Mail* ran a daily piece on his efforts to lure the monster from its lair and to bag the beast. And within just two days, the headlines announced he had found unusual footprints on the shoreline. A cast was sent to the British Museum for identification and the Scots were reveling in the global attention their country was receiving. But the following week they were hanging their heads in shame when the cast proved to be the imprint of a stuffed hippopotamus foot, probably an umbrella stand from some local hostelry or tavern. Weatherall denied any mischief making, and it was never proved whether it had been hunter or hoaxer who had laid the false tracks.

The two most compelling photographs of the "monster" are world famous. One depicts a creature with a long grayish neck

that tapers into an eerie thin head rising out of the water, followed by two humps. Roy Chapman Andrews, the American explorer and director of the American Museum of Natural History upon whom Indiana Jones was based, went on record in 1935 arguing that he had seen the original picture and that it had been "retouched" by newspaper artists before being published. He firmly stated the original picture was of the dorsal fin of a killer whale.

Most other experts disagree. As do I: to my mind, it is clearly the trunk of an elephant, with the first hump being the head and the second its back, almost certainly one of Bertram Mills's, taken as the circus elephants swam in the loch. Hugh Gray was the photographer: "I immediately got my camera ready and snapped the object which was then two to three feet above the surface of the water. I did not see any head, for what I took to be the front parts were under the water, but there was considerable movement from what seemed to be the tail." This photograph has been declared genuine by photographic experts and shows no signs of tampering, unlike so many of the others. And that is because, in my view, it is a genuine photograph—of a genuine elephant. No retouching required.

But the best-known photograph is the one taken by surgeon Robert Kenneth Wilson on April 19, 1934. Indeed, it must be one of the most instantly recognizable pictures ever taken. From a distance of two hundred yards, what has come to be known as the "surgeon's photograph" shows a gray "trunk" of around four feet protruding from the water with a hump directly behind it and clear disturbance in the water around. Once developed and declared genuine, the picture was bought and published by the *Daily Mail* and the Loch Ness Monster industry was properly born.

Curiously enough, when asked what he thought he had seen, Wilson claimed to have been too busy setting up his camera to take proper note, but thought there was certainly something strange in the loch. The next question then should have been: "Why didn't you wait around for a while to see if it returned?" If he had, he might well have seen the elephant surfacing, as it would have had to sooner or later. Then again, perhaps he did, but greed rather than valor influenced the better part of his discretion.

As recently as March 2006, Neil Clark, the curator of paleontology at the Hunterian Museum in Glasgow, has stated (thus confirming something I have believed for many years): "It is quite possible that people not used to seeing a swimming elephant—the vast bulk of the animal is submerged, with only a thick trunk and a couple of humps visible—thought they saw a monster." Dr. Clark also notes that most sightings came around the time of Bertram Mills's reward offer for evidence of the monster. He himself believes that most of the other sightings can probably be explained away by floating logs or unusual waves. But just as it seemed the eminent professor was about to finally blow the Loch Ness Monster out of the water, so to speak, he was asked by the BBC whether he believed there was a large creature living in the loch. To which he responded: "I believe there is something alive in Loch Ness." And he's not wrong, is he? There must be "something" alive in the loch; in fact there are lots of living things swimming around in it. But at least he didn't go on to say it was a fifteen-hundred-year-old sea monster, which it would have to be, as that is the premise upon which this whole story has been constructed.

But to be fair to Dr. Clark, the Loch Ness Monster is big business for Scotland. Consultants have estimated it to be worth in the region of £50 million per annum and rising. More than five hundred thousand tourists travel to the area every year in the hope of sighting the beast, despite Bertram Mills's reward expiring with him. Some claim the industry has even created twenty-five hundred new jobs. And the monster-spotting tour comes in at £15 a head. Dr. Clark would not be popular in his home country if he finally dispelled the myth many love and even more rely upon.

Since the elephant-heavy 1930s there have been dozens of sightings of objects of varying shapes and sizes. Even if paddling pachyderms are no longer the likeliest explanation, other theories are plausible. Loch Ness is actually a sea lake, fed from the Moray Firth in the North Sea via the River Ness. Furthermore, the Moray Firth is one of the areas of British seawater most frequented by porpoises, dolphins, and whales. Indeed, seals and dolphins have been filmed in the loch many times. If the mind wants to see a monster, three partly submerged dolphins swimming in a row could easily provide the illusion of a thirty-foot, three-humped creature in the gathering gloom—especially after a few drams of the local malt. I myself have encountered a few three-humped monsters after a lively evening out before now.

The BBC has used sonar and satellite imagery to scan every inch of the loch and found "no trace of any large animal living there." But, as has always been the case with myths, legends, and fables, while it is possible to prove the positive by producing irrefutable evidence, it is never possible to prove the opposite argument.

We could dam Loch Ness and drain it. We would then be able to take everybody still perpetuating the myth down into this vast new dry valley and show them every nook, cave, and rock cluster, but still the hard-core believers would reply: "Ah, but Nessie may well be out in the North Sea at the moment just limbering up for another appearance." But of course that is not the reason at all. Everyone from Columba (who told that miraculous story, embroidered or otherwise, which led to his canonization) onward has profited from retelling the tall tale of Loch Ness. The only surprise is that so many people have, and still do, strongly believe there is an unidentified prehistoric monster living in a Scottish loch. Some argue that it is a historical fact; I know it's just a hysterical one. I'm here to inform you, kids—there is no such thing as the Loch Ness Monster. Just don't tell anyone it was me who told you.

Will the Real Paul McCartney
Please Stand Up

Did the famous ex-Beatle really die
in a car crash back in 1966?

On October 12, 1969, Tom Zarski rang "Uncle" Russ Gibb's radio show on WKNR-FM in Dearborn, Michigan, and announced that Paul McCartney had been killed in an accident in November 1966 and the Beatles had drafted in a look-alike to keep the band fully functioning. He backed up his argument with several pieces of credible circumstantial evidence, including the decision by the band in 1967 to stop playing live in order to concentrate on their studio recordings and film work.

Russ Gibb was so intrigued by the story that he then spent two hours on air mulling over the clues and playing Beatles records. When one caller urged him to play "Revolution 9" (from the White Album) backward, Gibb was amazed to find he could distinctly make out the words "Turn me on, dead man" through his headphones. Even though Zarski had pointed out that he didn't actually believe Paul McCartney was dead, he was just interested in the theory, by the end of the program networks across the United States were discussing the mysterious death of one of the world's most famous rock stars and the events surrounding his demise. Hundreds of news journalists promptly flew to London and interviewed as many of the conspiracy theorists as they could find, and from the reports that followed the only certainty is that many of them were experimenting with LSD, as none of it made much sense at all.

The story ran that on the evening of Tuesday, November 8, 1966, Paul McCartney and John Lennon were working late into the night on the Beatles' upcoming album *Sgt. Pepper's Lonely Hearts Club Band,* when a row developed over recording techniques and McCartney stormed out of the studio. Furious, he sped off in his Aston Martin and smashed into a van, dying instantly. The resulting fire prevented the coroner from positively identifying the body, but the remaining band members were left in no doubt at all that McCartney had not survived. Another caller to Russ Gibb's show claimed that McCartney had picked up a hitchhiker that night. When the girl realized who he was, she had suddenly screamed and lunged at her hero, causing him to crash into the van. Neither the hitchhiker nor the other driver was ever seen or heard from again.

The public mourned as shock set in, but there was one unavoidable question: If McCartney had died in 1966, who was the man who looked like Paul and who had been hanging out with the Beatles ever since? The explanation ran that Beatles manager Brian Epstein was so horrified at the thought of the world's most successful band breaking up that he held secret auditions and persuaded John, George, and Ringo to have all their photographs taken with a stand-in to keep the public unaware of the accident. When Epstein died only nine months later, after a battle with depression and drug abuse, his untimely demise was cited as another piece of evidence. It was said that he just couldn't come to terms with the loss of McCartney. The Paul-is-dead mystery was also conveniently used to explain McCartney's sudden split from his long-term fiancée, Jane Asher (because McCartney stand-in William Shears Campbell didn't like her) and that his new relationship with Linda Eastman (later Mc-Cartney) was Campbell's real love interest.

Another piece of supposedly compelling evidence is that for several years the other three Beatles had wanted to stop playing live shows because the audiences were screaming so loudly they couldn't hear anything, but McCartney had resisted. With Paul gone, the remaining three could do as they pleased—indeed, the Beatles had last performed live on August 29, 1966, at Candlestick Park in San Francisco, and played no more live concerts after that. Conspiracy theorists nodded and agreed that it all made perfect sense, while others, including the Beatles, laughed it off as a ridiculous urban legend.

And still the story continued. One American radio announcer had photographs of the singer before and after November 1966 scientifically compared and found there were obvious differences, one being that the nose was of a different length. A doctor from the University of Miami analyzed voice recordings and concluded publicly that the recordings prior to August 1966 were different from those recorded afterward. Paul McCartney, he claimed, did not sing on Beatles records after August 1966.

By now fans all over the world were beginning to look for their own clues in Beatles music and album covers, and the clues turned up in spades. Here then are some of them, and the evidence seemingly pointing to the fact that Paul McCartney was dead.

Sgt. Pepper's was the first album the Beatles released after the supposed accident, after recording began on December 6, 1966. When it reached the shops in June 1967, nobody noticed anything unusual about the artwork in connection with the Paul McCartney mystery, but in 1969 conspiracy theorists were able to detect a range of coded references to Paul's demise. For a start, the band appear to be standing at a graveside complete with flowers and wreaths. They are surrounded by famous personalities, who could be mourners, and one of them is holding an open hand above McCartney's head, an open hand said to be a traditional Eastern symbol for death. The theorists looked closer and concluded that the yellow flowers at the foot of the picture are arranged in the shape of a left-handed bass guitar, Paul's instrument, and one of the four strings is missing, signifying his absence. Under the doll's arm on the right-hand side there appears to be a bloodstained driving glove; the doll itself has a head wound similar to the one Paul was supposed to have died from; and the figure of Paul is wearing a badge on his sleeve on the inside cover bearing the letters OPD, standing for "officially pronounced dead."

The open-palm gesture actually also appears on the front cover of *Revolver*, twice in the *Magical Mystery Tour* booklet, twice in the *Magical Mystery Tour* film, and twice on the cover of the original *Yellow Submarine* sleeve, but, in reality, none of it means anything at all. There is no such gesture in Indian culture symbolizing death. The badge Paul is wearing on the inside sleeve does not read "OPD"; it has the initials OPP on it. The badge was in fact given to McCartney when he visited the Ontario Provincial Police in Canada during the Beatles' world tour in 1965.

A statue of Kali, a Hindu goddess, is also featured on the front cover of the *Sgt. Pepper's* album, which the theorists maintain represents rebirth and regeneration, hinting that one of the Beatles has been reborn, or replaced. But Kali, from which the name of Calcutta is believed to derive, has traditionally been a figure of annihilation, representing the destructive power of time (*kala* being the Sanskrit word for "time").

Also, the O-shaped arrangement of flowers at the end of the band's name has caused some theorists to speculate that the whole thing reads "BE AT LESO" instead of "BEATLES." This was taken as a sign that Paul was buried at Leso, the Greek island the band had supposedly bought. But none of the Beatles had bought a Greek island, and there is no such place as Leso.

There are many more pieces of "convincing" evidence. I've just picked out some of my favorites.

The Beatles all grew mustaches at the time to help mask a scar on the lip of McCartney stand-in William Shears Campbell.

In fact McCartney did grow a mustache for *Sgt. Pepper's,* as he was unable to shave at the time. Paul had fallen off his scooter on his way to visit his aunt and split his lip on the pavement, making it too painful to shave. He also lost a front tooth in the accident, which explains why he appears in the "Rain" and "Paperback

Writer" promo videos missing one of his teeth. The accident also explains the scars seen during the *White Album* photograph sessions.

The license plate on the VW Beetle shown on the Abbey Road cover reads LMW 281F, taken to mean Paul would have been 28 "IF" he had survived.

But Paul would have been only twenty-seven, and the VW Beetle had nothing to do with anyone at Abbey Road. The director of the photo sessions tried to have it towed away, but the police took too long to arrive, so they went ahead with the picture anyway, leaving it in the shot.

McCartney is wearing no shoes in the Abbey Road *photograph*.

His explanation was: "It was a hot day and I wanted to take my shoes off, to look slightly different from the others. That's all that was about. Now people can tell me apart from the others." But the conspiracy theorists swore that the picture had been set up to look like a funeral march, with him as the corpse.

On the records *Rubber Soul, Yesterday and Today, Help!* and *Revolver* there were said to be many more clues. The song "I'm Looking Through You" on *Rubber Soul* was thought to be about discovering that McCartney had been replaced. Some fans took these blatant "clues" as hard evidence, while others quickly realized all of those records were made prior to November 9, 1966, and could not possibly have anything to do with the supposed accident.

With hysteria mounting, however, even the thinnest clue

came to look like definite evidence. In the lyrics to "I Am the Walrus," the line "stupid bloody Tuesday" is taken by some to be John Lennon's reference to the day of the accident that claimed his bandmate. But when it was pointed out that the alleged accident was supposed to have happened on a Wednesday morning, conspiracy theorists then claimed it was the Tuesday night that the two of them had fallen out before McCartney had stormed off to his death. Some believed it, while others dismissed it as an already thin lead being stretched even thinner. But then came the line "waiting for the van to come," a supposed reference to the ambulance, and "goo goo ga joob"—apparently Humpty Dumpty's last words before he fell off that wall and bashed his head in.

The Beatles themselves very quickly became very irritated by all the speculation. And it was not long before the band, aware that every lyric and photo shoot was now being studied, began to play up to the hysteria. After writing one complicated and seemingly meaningless song called "Glass Onion," Lennon remarked, "Let the f**kers work that one out." But he included the lines "Well here's another clue for you all / The walrus was Paul." In no time at all, people were announcing that the walrus was a symbol of death in some cultures, and Lennon despaired. It wasn't much fun being a Beatle anymore, and the band broke up soon afterward.

So—to sum up—if the real Paul McCartney had died in his Aston Martin in 1966, and a replacement had been found in time for the photo shoots for the next album, then imagine the string of coincidences that needed to have taken place. For a start he had to look and sound just like Paul. Then he had to convince Linda or, if she was in on the plot, she had to like him enough to stay married to him for the next thirty years. And he would have had to learn how to play guitar left-handed, which is even less likely, I can assure you. John Lennon would have to have been fooled too, as it is unlikely he would have wanted to share songwriting credits and royalties with a stranger for the last three years of Beatles recordings, especially as Epstein wasn't there to tell him to. And most of all, for the look-alike to have written and recorded songs of a McCartney standard for over thirty years would be hard to imagine.

Hang on a minute, I have just remembered "The Frog Chorus" and "Mull of Kintyre," and so my argument is beginning to wear thin. Perhaps Zarski was right in the first place—there must be an impostor . . .

The Magnetic Strip

....

*The strange story of Germany's most
dangerous stretch of road*

There are hundreds of unexplained mysteries from every corner of the planet involving cars, drivers, hitchhikers, car theft, and abduction. One of the most unusual occurred soon after a new section of the autobahn in Germany was opened to traffic between Bremen and Bremerhaven in 1929. During the first year alone no fewer than a hundred cars crashed or came off the autobahn, but the accidents were all happening in exactly the same place, very close to kilometer marker number 239. On one particular day, September 7, 1930, nine separate accidents took place adjacent to the marker post, in each of which all vehicles were destroyed.

There appeared to be no explanation for the accidents, as the stretch of road in question was flat and straight and no hazards had been reported. And that day in September had been particularly fine and sunny. However, survivors told police that when they approached the marker they had felt a sensation in their stomachs as if they had crossed a humpback bridge at speed, and a "strange force then took over the steering and threw [their] car off the road."

German police were flummoxed until a local water diviner, Carl Wehrs, suggested that a powerful magnetic force caused by an underground stream might have been the reason for the mysterious accidents. Accompanied by witnesses, he walked with a

steel divining rod toward the marker. He was about ten feet away when, all of a sudden, the rod was ripped from his grasp, the sheer force of it spinning his body around 360 degrees, like an Olympic hammer thrower.

Wehrs's solution to the problem was to bury a box of copper next to the marker stone, and the accidents immediately stopped. To further test his theory, he later dug the box back up, and the first three cars to pass by all crashed. Once the box was reburied the marker post was removed, the area was sprinkled with holy water, and the accidents ceased and have never recurred.

Whatever Happened to the Crew of the *Mary Celeste*?

························,·························

The **Mary Celeste** *was a ghost ship found off the coast of Portugal in 1872. Why she had been abandoned has been the subject of endless speculation ever since.*

One calm, quiet afternoon in December 1872, seaman John Johnson peered through his telescope from the deck of the *Dei Gratia* ("by the grace of God" in English). Alarmed by what he had seen, he shouted down for the second mate, John Wright, to join him, and the two men stared at the ship sailing erratically on the horizon. They then summoned the captain, David Reed Morehouse, and the first mate, Oliver Deveau. Morehouse at once recognized the *Mary Celeste*, which had put to sea from New York only seven days before the *Dei Gratia*. Despite the absence of distress signals, Morehouse knew something had to be wrong—no one appeared to be guiding the vessel—so he steered his ship closer. After two hours, Morehouse concluded the *Mary Celeste* was drifting, so he dispatched Deveau and some deckhands in a small boat to investigate, and one of the most puzzling sea mysteries of all time began to unfold, for the brigantine was completely deserted.

It was later recorded—although not by Deveau himself, who kept his information for the later inquest he knew he would have to attend—that the boarding party came upon mugs of tea and a half-eaten meal left out on the table, and a fat ship's cat fast asleep on a locker. Mysterious cuts had been made in part of the railing, some strange slits had been cut into the deck, and a bloodstained sword was discovered under the captain's bed. Two small hatches to the cargo hold were open, although the main one was secure, and nine of the 1,701 barrels of American alcohol were empty. A spool of thread was balanced on a sewing machine and, given the slightest movement, would clearly have rolled off if the sea hadn't been so calm. A clock was turning backward and the compass had been broken, but there were no signs of a violent struggle and, even more mysteriously, no sign of Captain Briggs, his wife, daughter, the single passenger, or any of the seven-man crew. Curiously, the vessel's sextant, navigation book, chronometer, ship's register, and other papers were all missing, while the captain's log lay open and ready for use upon his desk. It appeared that the people on board the *Mary Celeste* had simply vanished in the middle of eating their breakfast, never to be seen again. This is the story that became the accepted version of events, but as we delve into the truth of the tale we will try to find out what really happened and how the legend has grown to become one of the greatest sea mysteries of all time.

Following the discovery of the ghost ship, people's imaginations were working overtime. *The Boston Post* reported on February 24, 1873, that "it is now believed that the brigantine *Mary Celeste* was seized by pirates in the latter part of November, and that the captain and his wife have been murdered." Two days later, *The New York Times* concluded that "the brig's officers are believed to have been murdered at sea." Ever since then, speculation about the crew's sudden disappearance has been the subject of many a seafaring yarn, with stories of mutiny, giant whales, sea monsters, alien abduction, and much more, while the truth of what happened to the people on board the doomed ship, discovered halfway between the Azores and the Portuguese coast on that calm December afternoon, has remained a mystery.

Frederick Solly Flood was the attorney general for Gibraltar, where the *Mary Celeste* had been taken by Morehouse and his crew, and the advocate general for the British Admiralty Court. An arrogant, excitable character, infamous for his snap decisions, he had lost his son's entire inheritance on a horse called the Colonel in the 1848 Epsom Derby. At the inquest into the *Mary Celeste,* Flood decided that the crew must have broken into the cargo hold and drunk the nine barrels of liquor before murdering the captain and his wife and abandoning ship. He had to rethink his ideas after it was pointed out that the *Mary Celeste*'s cargo was of denatured alcohol, a mixture of ethanol and methanol similar to methylated spirits, and more likely to kill than to intoxicate.

Unabashed, Flood revised his conclusion to suggest a conspiracy between the two captains, who knew each other, to defraud the *Mary Celeste*'s owners. According to this theory, Briggs had killed his crew just before Morehouse was due to intercept the *Mary Celeste* and then stowed away with his family on the *Dei Gratia* while Morehouse claimed the salvage rights to the *Mary Celeste* and the two scurrilous captains split the money. It was then pointed out to the hapless attorney general that Briggs part-owned the ship himself and that the entire salvage money would have been less than his original investment. Solly Flood went back to the drawing board and decided that if Briggs hadn't been involved, then Morehouse must have killed the entire crew to gain salvage rights to the ship himself. Eventually, after many months of slander, the Admiralty stepped in and exonerated Morehouse of all responsibility, compensating him and his crew. Oliver Deveau must have read in despair what had been attributed to him by the newspapers, to which a spiteful Flood had been quick to leak details of the case.

Other theories were also dismissed, since giant sea monsters, despite a penchant for sailors, were not known for taking a ship's papers and navigational instruments, nor were the aliens who had apparently abducted every living being on board except the cat. Neither were they picked off the deck one by one by a giant sea squid, nor blown into the sea by a passing whale that sneezed, and most clear-thinking people have ruled out any connection with the Bermuda Triangle (see "Try to See It from My Angle: The Bermuda Triangle," page 12), as the *Mary Celeste*'s path didn't cross it. Piracy was also ruled out, as nothing of value had been stolen, and mutiny was considered unlikely, as the small crew of professional and disciplined sailors were on the short voyage voluntarily and Captain Briggs himself was known to be well liked by his men. In March 1873, the court finally had to admit they were unable to determine the reason why Captain Briggs had abandoned the *Mary Celeste*, a conclusion that caused a sensation as it was the first time in history a nautical inquest had failed to find a satisfactory explanation.

It was Solly Flood's rantings in court that alerted the English media to the mystery of the *Mary Celeste*. When news reached London, a certain young doctor took a keen interest in the reports, using them in a short story, "J. Habakuk Jephson's Statement." The yarn, published in January 1884 by the prestigious *Cornhill Magazine*, featured a mystery boat called *Marie Celeste*, not *Mary Celeste*, captained by a man called Tibbs, not Briggs. Many features of the fictional account are close to the true story of the *Mary Celeste*. Equally, many details—such as the half-eaten breakfast, and the abandoned boat in perfect condition floating serenely on still waters—were a figment of the writer's imagination. And as the imagination belonged to the young

Arthur Conan Doyle (who also crops up in "Fairies at the Bottom of the Garden," page 101, and "The Spine-chilling Tale of the Chase Vault," page 39), the creator of Sherlock Holmes, it was extremely convincing. With his appealing mixture of fact and fiction, Conan Doyle had inadvertently created a mystery that would occupy thousands of minds over the next century and provoke endless hours of debate.

Just when the conspiracy theories surrounding the *Mary* (not *Marie*) *Celeste* were beginning to die down, an interesting new lead emerged. In 1913, Howard Linford came across some old papers of Abel Fosdyk, a friend of his who had recently died. Among them was what claimed to be an eyewitness account of what had happened to the captain and crew of the *Mary Celeste.* According to this account, Abel Fosdyk, due to unfortunate circumstances, had had to leave America in a hurry and had persuaded his good friend Captain Briggs to stow him away on the *Mary Celeste.* He also describes how Briggs had asked a carpenter to install a new deck-level on board so that his wife and daughter would have a viewing platform away from the dangers of a working ship's deck. Fosdyk then tells how Briggs, while at sea, became involved in a good-natured argument with two of the crew about how well a man could swim while fully clothed and to conclude the matter all three jumped into the calm water for a race. Unfortunately, they were then attacked by passing sharks. When the rest of the crew raced up onto the new temporary deck to see what the commotion was, it promptly collapsed, throwing everybody to the sharks. Everyone apart from Fosdyk himself, that is, who clung to the platform, which drifted to the coast of Africa where he was saved. According to Fosdyk, he had been unable to tell the story during his lifetime for fear of being recognized and hauled back to America.

However, Fosdyk had gotten many of his facts about the ship and crew wrong. He claimed the crew were entirely English when in fact the crew list confirms four were German. Also, he described the *Mary Celeste* as a vessel of six hundred tons when in reality it was less than half that size. Finally, it is highly unlikely that Briggs, a responsible sea captain, would jump fully clothed into the sea with two of his crew, leaving the rest of his men, his wife, and his two-year-old daughter on board to fend for themselves should the three swimmers run into trouble. Especially as, given the set of the rigging when the boat was discovered deserted by the *Dei Gratia*, it must have been sailing at a speed of several knots at the time, leaving the swimmers far behind. Whether Fosdyk invented the story and left it to be discovered among his papers upon his death, or whether his friend Howard Linford created the myth, is unknown.

Nevertheless, when *The Strand Magazine* published the papers in 1913, they raised more questions about the mystery than they answered. Then, in the late 1920s, in *Chambers's Journal,* a young reporter by the name of Lee Kaye interviewed John Pemberton, another alleged sole survivor of the *Mary Celeste* claiming to be able to reveal the shocking truth of what had really happened to the captain and crew. The public demanded to know more and the press eventually tracked Pemberton down and published the story alongside a photograph of the old sailor. Lawrence Keating turned the story into a book, *The Great Mary Celeste Hoax* (1929). The book was a worldwide bestseller until it was revealed that the journalist Lee Kaye, the sailor John Pemberton, and the author Lawrence Keating were all one and the same. The photograph of Pemberton that Keating had given the press was of his own father.

But setting all the hoaxes and theories aside, what really did happen to the *Mary Celeste*? Let's consider the evidence in a bit more detail.

In 1861, the first ship to emerge from the yard of Joshua Dewis shipbuilders on Spencer Island, Nova Scotia, was chris-

tened the *Amazon*. Launched as the American Civil War was gathering pace, she proved to be trouble right from the start. Her first captain, Robert McLellan, died before the ship went anywhere. Her second captain, John Nutting Parker, sailed her into a weir in Maine and during the subsequent repairs she caught fire. The ship passed through many hands with equal bad luck before being bought by J. H. Winchester & Co. of New York for $2,500 during 1871. Captain Benjamin Spooner Briggs bought a one-third share in the boat, intended to be his retirement fund. Briggs was born on April 24, 1835, in the town of Wareham, Massachusetts, and was a man of strict religious beliefs and a dedicated teetotaler; he was described as "of the highest character as a Christian and an intelligent and active shipmaster." After a $14,500 refit, the ship reemerged in New York's East River proudly bearing a new, hopefully luckier name. The rechristened *Mary Celeste* was ready for her maiden voyage.

In 1872, Briggs prepared to take his new ship to Genoa with

a cargo of denatured alcohol (intended for use by the Italians to fortify their wines). He enlisted his first crew, engaging Albert Richardson, a Civil War veteran who had served twice before with Briggs, as first mate. Second mate Andrew Gilling and steward Edward William Head were also of solid and reliable reputation. The four ordinary seamen were all German, two being brothers who had recently survived a shipwreck that had destroyed all of their possessions.

On Saturday, November 2, 1872, after the barrels of alcohol had been loaded and made secure, Captain Briggs is known to have dined with his old friend Captain David Morehouse, skipper of the *Dei Gratia,* who had a cargo of petroleum to transport to Gibraltar a little over a week later. The two ships would be taking an almost identical route across the Atlantic, although the two men did not expect to see each other again before they returned to New York. As the weather was particularly stormy in the Atlantic, Captain Briggs was forced to wait before he risked venturing out on the open sea, and he finally set sail on the afternoon of November 7.

According to the captain's log, later found in Briggs's cabin, the voyage was uneventful until the last entry recorded on November 25, which noted that the ship had reached St. Mary's Island (now called Santa Maria) in the Azores. At that time the weather was deteriorating badly and the ship had been speeding along on a northeasterly wind toward the Azores. Captain Morehouse later testified that these strong winds soon turned into a torrential storm with gale-force gusts. This may explain why Captain Briggs had sailed the *Mary Celeste* to the north of St. Mary's Island in the hope of finding some relief from the harsh weather. Nothing else is known of the fate of the *Mary Celeste* or her crew, and nothing is known of their whereabouts between November 25 and December 4, when the crew of *Dei Gratia* found the *Mary Celeste* adrift halfway between the Azores and the Portuguese coastline. However, the official evidence provided at the subsequent inquiry in Gibraltar provides plenty of clues.

Oliver Deveau, the seaman in charge of the boarding party, found no lifeboat aboard the *Mary Celeste,* despite the generally accepted belief that the lifeboat remained secured to the deck, which added to the intrigue. There may have even been two lifeboats on board when the ship left New York. He found that the front and rear cargo hatches had been removed and placed on the deck with sounding rods nearby, suggesting the hold was being measured for water intake, or perhaps being aired, at the time the crew disappeared. Only one pump was working, and there was a great deal of standing water between the decks, with another three and a half feet in the hold. However, despite his noting that the ship was a "thoroughly wet mess with the captain's bed soaked through and not fit to sleep in," Deveau declared the ship seaworthy and sound enough to sail around the world in his view.

He also recorded that although some of the rigging and the foresails had been lost, they had not been lashed properly and might have come adrift at any point. The jib, the fore-topmast staysail, and the fore lower topsail were set and the rest of the sails were all furled, suggesting the crew were already making ready to raise anchor and were in the process of setting the sails at the time they disappeared. There was ample freshwater and food in the galley, but curiously the heavy iron stove had been knocked out of its retaining chocks and was lying upturned on the deck.

A large water barrel, usually held in place, was loose and rolling free, and the steering wheel had not been lashed into position (normal procedure when abandoning ship). There were strange cuts on the rail and hatch where the lifeboat tied to the main hatch had been axed free, rather than untied, and part of the railing had been hacked away to allow the lifeboat to be launched quickly. The apparently bloodstained sword had, in fact, been cleaned with lime, which had oxidized the blade red. Solly Flood had known this, but chose to withhold that information from the court. Finally, and mysteriously, the ship was missing the American flag so proudly displayed as she left New York. It is clear that the *Mary Celeste* was abandoned in great haste, but the question is why Captain Briggs would desert a perfectly good ship for a small lifeboat. What happened on board to cause an experienced captain and crew to jump off the ship and into a tiny lifeboat, where they would be in far greater danger, when it must have been obvious the *Mary Celeste* was in no danger of sinking?

James H. Winchester, part owner of the ship, suggested at the time that the cargo of raw alcohol could have given off powerful fumes and that this might have gathered in the hold and formed an explosive cocktail. He speculated that a spark caused by the metal strips reinforcing the barrels rubbing against each other could have ignited this, or that perhaps a naked flame used to inspect the hold could have caused a vapor flash, not strong enough to create any fire damage but frightening enough to suggest to the captain and crew that the whole cargo was about to explode. Furthermore, Oliver Deveau stated at the salvage hearing that he thought something had panicked the crew into believing the ship was about to sink and so they had taken to the lifeboat. The theory fits the evidence almost perfectly, but does not explain all the water found on board or the heavy water butt and iron stove being knocked out of their secure fastenings. The clock with backward-rotating hands was not as mysterious as first thought after Deveau explained that it had been placed upside down, evidently by mistake.

A more recent theory, though, has at last provided a far more credible explanation as to what happened on board that morning—one that even the ingenious Conan Doyle would not have dreamed up. Not a waterspout or tornado at sea, but a seaquake (see also "The Mysterious Disappearance of the Lighthouse Keepers of Eilean Mor," page 90). Could an offshore earthquake finally provide the answer mystery lovers have spent over a hundred and forty years searching for? The United States Naval Research Laboratory has recorded that a major seaquake has occurred within a short distance of the island of Santa Maria every year since records began. On November 1, 1755, just over a century before the *Mary Celeste* was found, an earthquake along the same fault line destroyed the port of Lisbon in Portugal. Falling buildings and the subsequent tsunami killed approximately one hundred thousand people. The section of ocean bed known as the East Azores Fracture Zone is thirty to forty miles southwest of Santa Maria, while approximately twenty miles northeast of the island lurks the Gloria Fault. This area is one of the seaquake capitals of the world, and the *Mary Celeste* was berthed right on top of it on the morning of November 25, 1872.

Dr. Lowell Whiteside, a leading American geophysicist, was asked in an interview to confirm if a seaquake might have taken place near Santa Maria on November 25, 1872. Whiteside started by pointing out that as seismological instruments were not then available, the only earthquakes recorded would have been the ones that were strong enough to be obvious, or in which there had been survivors. He went on to confirm: "The Azores is a highly seismic region and earthquakes often occur; usually they are of moderate to large size." He then added: "An 8.5 magnitude seaquake did occur in the Azores in late December 1872 and that was recorded. This was the largest in the area for over one hundred years and it is probable that many large foreshocks and aftershocks would have occurred locally within a month either side of this event." The 8.5 magnitude earthquake in December 1872 was reported on every island of the Azores, such was its scale, but foreshocks and aftershocks would not necessarily have made the news and therefore would not have been recorded.

Newly armed with evidence of a major earthquake and "highly probable" foreshocks at exactly the time *Mary Celeste* was known to be in the area, investigators appeared to have hit upon a perfect solution to the mystery. A seaquake would cause a vessel the size of the *Mary Celeste* to shudder violently and, when directly over the fault line, to bounce up and down as the waves were forced vertically toward the surface. This would explain the topsails being partly set, as the two crew members high in the rigging would certainly have been thrown off and into the sea. Other sailors have witnessed craft caught in a seaquake and report that at times the ship would be completely surrounded by a wall of water, explaining why *Mary Celeste* was wet inside and also why the captain's bed was unmade as well as soaked through. No doubt Captain Briggs was thrown awake from his bed to find his crew panicking at the commotion that would have appeared without warning and from a previously calm sea.

The violent bucking would have dislodged the heavy stove and water butt, and sent hot ash and smoke around the galley. The thundering noise would have been terrifying and the whole event something even an experienced crew like the one on *Mary Celeste* would never have been through. Nine barrels of alcohol could easily have been damaged in the process, causing nearly five hundred gallons of pure alcohol to spill into the hold. Suspecting damage to the barrels, the crew may have removed the hatch to the hold to investigate. As the alcohol fumes issued from below, they could have been ignited, either by the stove coals or metal sparks from the hatch lid, creating a blue vapor flash that wouldn't necessarily have resulted in fire damage. Any amateur investigator can re-create this effect by removing the lid of an empty rum or brandy bottle and dropping in a lighted match. The resulting vapor flash will often force the match straight back out. Placing rolled-up paper balls in the bottle will also prove that no burn damage is caused by such an event. Old sailors called this trick "igniting the genie." But if you want to try it at home, then do it outside—and don't set fire to your mum's curtains.

Under the circumstances, it is easy to see how Captain Briggs and his crew could have feared that the cargo was about to explode and think that they should abandon ship immediately. They may even have believed the volatile alcohol rather than a seaquake—something of which comparatively little was known at the time—was in some way responsible for the ship's unnatural behavior. Given the perceived threat, Briggs would undoubtedly have evacuated his family and crew to a safe distance in the lifeboat, and this was obviously done in great haste, the captain only stopping to gather up his navigational instruments and the ship's papers and registration documents. Whether deliberately or by accident, the lifeboat was not secured to the mother ship by a length of rope, as would have been normal in the case of evacuation.

But the drama would have soon been over and the confused crew may well have sat in the lifeboat watching the *Mary Celeste*, with her partly set sails, calm, afloat, and in no apparent danger. The captain would then have had a big decision to make: either head in the lifeboat to Santa Maria Island and explain why he had abandoned a perfectly seaworthy ship with its valuable cargo on the evidence of some strange bouncing motions and a few ghostly blue flashes, or start after his ship in the hope of catching up with her and regaining command. What has been rarely connected to this story is the fact that in May the following year, fishermen discovered a badly damaged raft washed ashore in Asturias in Spain, with five badly decomposing bodies and an

American flag on board. For some investigators this proves Captain Briggs attempted to catch up with his ship in the lifeboat, with tragic consequences.

Without the inventive fiction of Arthur Conan Doyle, with his half-eaten breakfast, sleeping cats, or delicately balanced reels of cotton, the story of the *Mary Celeste* is not as ghostly as it seems. The theory that she was caught up in a frightening seaquake and abandoned would seem to silence any conjecture about supernatural goings-on. No doubt, however, various storytellers or creative Hollywood minds will bring new theories to our attention in the continuing debate about the fate of *Mary Celeste*'s crew. Perhaps they will reintroduce aliens, hungry sea monsters, or a giant man-eating bird of prey, but for this investigator the answer is found in the violent seaquake that caused Captain Briggs to abandon ship and then drift to his death with his wife, baby daughter, and remaining crew.

Although the most famous, the *Mary Celeste* is by no means the only ship to have been found deserted at sea. In April 1849, the Dutch schooner *Hermania* was discovered floating off the Cornish coast, near the Eddystone lighthouse, without her mast. In this case, the lifeboat was still firmly lashed to the deck and all personal belongings were in the cabins. However, the captain, his wife and daughter, and all the crew members were never seen again. Six years later another ship, the *Marathon*, was found adrift with no hands on deck and in perfect condition.

So what became of the most famous ghost ship in history? After being released by the authorities in Gibraltar, she returned to New York, where J. H. Winchester promptly sold her. On January 3, 1885, she ran onto the razor-sharp rocks at Rochelais Bank in the Gulf of Gonâve and was wrecked. Unfortunately for her new owner, Gilman Parker, his insurance company decided to send an investigator to inspect the wreck before paying his claim for $30,000. The investigator found the cargo to have no value at all, made up as it was of cat food, old shoes, and other rubbish. It turned out that Parker had unloaded the small part of the cargo with value and then had set fire to the *Mary Celeste*.

Parker was promptly charged with fraud and criminal negligence, a crime punishable by death in 1885. Then a legal technicality forced prosecutors to withdraw the charges laid against Parker and his associates and they were released, but the *Mary Celeste* still exacted her revenge. Over the next eight months one of the three conspirators committed suicide, one went mad, and Parker himself was bankrupted and died in poverty. And so the story of the *Mary Celeste* ends, leaving us with not only one of the best-loved and most intriguing mysteries in seafaring history, but also one of the most tragic.

The Men Who Cheated Death

••

Two men they couldn't hang and one man
miraculously cured of his war wounds

As John Lee was being led from his cell in Exeter Prison in England on the morning of February 23, 1885, he had approximately two minutes left to live. He was making the condemned
man's walk to the gallows, having been convicted of the murder
of the elderly woman he worked for, Emma Ann Keyse, who had
been discovered with her throat cut and her head battered. Lee
had protested his innocence, but his criminal record, obvious
hatred of his employer, and lack of alibi had sealed his fate. As
his jailers led him to the scaffold, his arms were strapped behind
his back, a white hood was placed over his head, and a noose was
secured around his neck. His executioner asked if he had any
last words or confession to make, and when John Lee replied,
"No, drop away," the sheriff of Exeter gave the order to proceed.

But when the executioner, Mr. Berry, pulled the lever to the
trapdoors, nothing happened. They didn't open, and Lee remained standing, alive if not exactly well, in the same location.
Without removing the noose, the executioner's men shuffled
Lee to one side and tested the doors, and this time they opened
smoothly. Lee was then edged back into place, directly in the
middle of the trapdoors, and the lever was pulled for a third
time. Once again he stood on the unsecured trapdoors, just a
few feet away from eternity, waiting for them to open. But yet
again, they refused to do so.

The sheriff of Exeter then ordered the condemned man back to his cell while a full examination was made of the gallows. The hangman himself stood on the trap and held on to the rope with both hands. As soon as the lever was pulled, he fell through. John Lee was once more brought from the condemned man's cell, restrained, and placed into position in the center of the trapdoors. But once again, when the lever was pulled, the doors refused to budge. The gathered crowd of witnesses, including newspaper journalists, were by that time shivering with cold, as was the condemned man standing upon the unsupported trapdoors awaiting his rather prolonged fate.

John Lee was then led back to his cell while further tests were made to the trapdoors, which, each time, worked perfectly. Completely baffled, the sheriff then wrote to the home secretary in London for further instructions. Newspapers across the world reported the story and John Lee found fame as "the man they cannot hang." The home secretary, Sir William Harcourt, ordered a stay of execution, the unprecedented event was discussed in the House of Commons, and John Lee's death sentence was commuted to one of life imprisonment.

The official explanation given was that the damp weather had caused the trapdoors to swell and become jammed, despite all the tests confirming the mechanism worked smoothly. The mystery of the man they could not hang was never solved, and John Lee was eventually released from prison in 1907. He later married and lived quietly in London until his death in 1943, a full fifty-eight years longer than was expected when he made that early-morning walk at Exeter Prison all those years ago.

A few years after the failed execution of John Lee, in 1894, a young American farmer named Will Purvis was sentenced to be hanged on February 7, for the murder of a farm owner in Co-

lumbia, Mississippi. As he was secured upon the platform of the gallows, a priest read out the last rites and the lever was pulled. The trap opened with a crash and Purvis plunged through, emerging from below covered with dust but otherwise unharmed. The noose had somehow become untied and slipped from his neck. Shocked, yet undeterred, deputies led him straight back up the steps and once again restrained him, carefully checking that the knot was secure this time. But the crowd gathered below, of around three thousand people, was by then singing and shouting that Purvis had been reprieved by the highest power of all, the Lord Himself, and threatened to become unruly if the hanging went ahead.

Sheriff Irvin Magee, very wisely under the circumstances, had the murderer escorted back to his cell instead of attempting to carry on with the execution. Purvis's defense team made several appeals to have his death sentence commuted, but without success, and a new date for the execution was set, December 12, 1895. Purvis then escaped from jail and went into hiding, but when Mississippi elected a new governor who was sympathetic to the young man's plight, Purvis surrendered himself and immediately had his sentence reduced to one of life imprisonment. By then Purvis had become a statewide hero and received thousands of letters of support demanding that he be considered for a full and complete pardon. The new governor agreed, and in 1898 Will Purvis was a free man.

And it was just as well, because in 1917 one Joseph Beard announced on his deathbed that he was responsible for the murder of the farmer, and not Will Purvis after all. Other details were given that proved his story, and Will was finally exonerated. He had always protested his innocence and, as his death sentence had first been announced in 1893, had broken down in tears and cried out to his accusers: "I will live to see every last one of you dies!" When he finally died, peacefully and without assistance, on October 13, 1938, it was noted that the last of the jurors to have found him guilty of murder had himself passed away only three days earlier. Nobody could ever explain how, without help of a supernatural kind, the noose had managed to slip from his neck, allowing him to cheat certain death.

During the First World War, a soldier from Liverpool, Jack Traynor, was serving in the trenches when he was hit twice by enemy fire. The first bullet hit his head, smashing his skull, while the second bullet hit his right arm, severing vital nerves that even the most skilled surgeon of his day was unable to reconnect. Jack's skull injury refused to heal—indeed, doctors believed he would soon succumb to the wound—and he became virtually paralyzed in his damaged arm. Consequently, he was awarded a full disability pension. It is recorded that a few years after the war, in 1923, Jack began to suffer from severe bouts of epilepsy, as a result of his head wound, and lost the ability to walk. During that year he was taken on a religious pilgrimage to Lourdes in France, where he was lowered by his family into the supposedly healing waters. After a short ceremony he was taken back to the hospice he had become confined to and placed gently back into bed. However, four days later Jack awoke and sprang from his bed, miraculously made whole again. He then washed, shaved, and dressed himself, packed his bags, and walked out of the hospice, never to return.

When Jack arrived back home in England, he set up in business as a coal merchant, met a young lady, fell in love, got married, and fathered two healthy children. He lived a normal, happy life for the next twenty years until he sadly died, in 1943, of pneumonia. Jack's well-being and prosperity must have been all the greater since throughout this time the Ministry of Pensions had refused to believe he could have made such a recovery and continued to pay his disability pension in full. Nobody has ever been able to explain Jack Traynor's remarkable and mysterious recovery.

Not in the Mood:
The Real Glenn Miller Story

··

The famous bandleader vanished without a trace
en route to entertain Allied troops in 1944,
but what happened to him?

At the end of the 1930s, just as the Second World War was breaking out in Europe, Glenn Miller's band introduced America to
the new, unique style of brass-band music they had been working on for a number of years. It was a smooth, upbeat sound that
struck an instant chord both with the middle-aged and with
optimistic youth learning how to jive and swing.

Radio stations across America played Glenn Miller records all
the time, and Hollywood was quick to sign up the new star and
his band. Two films were released: *Sun Valley Serenade* in 1941
and *Orchestra Wives* in 1942. The Glenn Miller Orchestra were
the Beatles of their generation (or, for the younger reader,
Oasis; and if you're thinking of the Arctic Monkeys, then you
should be in bed by now). By early 1942, America had entered
the fray, joining the Allied forces in their efforts to repulse the
Nazis. Miller enlisted later that year, on October 7. On comple-
tion of his basic training, he transferred to the Army Air Corps;
his first military assignment was to gather another orchestra, the
Glenn Miller Army Air Force Band, with a brief to entertain Allied troops in Britain. He was delighted to be back in touch with
his old Hollywood friend David Niven, whose job it was to
arrange entertainment for the troops across Europe.

Eighteen months later, the D-Day landings signaled the start of the liberation of Europe, and by November 1944, Paris was finally free of German soldiers. Even though Allied bombers were still pouring across the English Channel on their way to tackle targets farther into Europe, the Parisian party was now in full swing. David Niven organized a six-week tour for the Glenn Miller Orchestra that was to begin in the French capital on December 16, 1944. The band was due to arrive on the sixteenth, but Miller wanted to travel earlier to attend what he called a "social engagement." Arrangements were duly made for him to fly from the airfield at Twinwood Farm near Bedford in a small American-built, propeller-driven craft called a Noorduyn Norseman that would be piloted by John R. Morgan. Lieutenant Don Haynes, a show-business agent drafted into the U.S. Army Air Force to manage the Glenn Miller Orchestra while on tour, drove his famous charge from London to RAF Milton Ernest Hall to prepare for his cross-Channel flight the following day. According to Haynes, John Morgan arrived in the Norseman at Twinwood Farm at 1:40 P.M., collected Miller, and, in spite of poor weather conditions, took off again at approximately 1:45 P.M. This was the last anyone saw of Glenn Miller: he had vanished from the world and into the history books.

The alarm was raised when he failed to meet up with Don Haynes and the band in Paris a day later. After a frantic search of the entire city's likely haunts, the Glenn Miller Orchestra had to play the show without their famous bandleader, announcing that "Major Miller cannot be with us tonight." Nobody ever saw him again, or at least could prove that they had. The puzzle began in earnest when, just three days later, the United States military announced his death, which was extraordinary in itself, given that in the confusion of a recently liberated France many people went missing for much longer periods, often "absent without leave" (AWOL).

The question was, Why would officials make such a final announcement so soon after the musician, albeit a world-famous one, simply failed to show up at a few concert performances? Pete Doherty does that all the time these days and nobody declares him dead as a result. It was a question Helen, Miller's wife, also asked, but not until over a year later, in February 1946, when Colonel Donnell wrote to inform her that her husband had been flying that day in a combat aircraft, not the Norseman, and that the plane had taken off from Abbots Ripton airfield near Huntingdon in Cambridgeshire, many miles from where Haynes had left Miller.

The mystery deepened when it was claimed that the flight had been bound for Bordeaux, far from Miller's intended destination. There was no explanation of how he would be traveling the remaining distance within France. In fact, no further information was given at all, and so speculation raged about whether Miller had lied about his movements to his friends and the rest of the band, changing his stated plans at the last minute, or had gone AWOL, or even that he had been shot down by enemy fire. A military cover-up seemed increasingly likely. Imagine that: the military might not be telling the truth about something!

After the war, John Edwards, a former RAF officer, set out to prove Miller *had* been on board the Norseman, for which all he needed was a copy of the official accident report from the National Personnel Records Center in St. Louis. But he drew a blank: that office maintained the records had been "lost in a fire," while the Department of Records in Washington, D.C., denied such a file had ever existed. Edwards's efforts to prove the absence of a military cover-up began to convince him that the reverse must be true.

What he now wanted to know was why. And when some documents were finally discovered, they were found to be written illegibly, the signatures blurred and indecipherable. This, strengthened by the fact that the military had initiated no search of any kind for the missing bandsman, began to fuel speculation that the U.S. government knew exactly what had happened to Glenn Miller and had known it immediately, hence the early announcement of his death. After all, imagine Oasis singer Liam Gallagher going missing on a morale-raising visit to troops in Iraq, there being no search for him, and the British government firmly announcing he was dead only three days later, but without producing a body. Furthermore, no records of what had happened to him would ever be released, while every government agency claimed to know nothing about it.

What *is* known is that the Norseman had crashed into the sea, as it was discovered by divers in 1985 six miles west of Le Touquet on the northern coast of France, but there was no evidence that Miller, or indeed anyone else, was on board at the time, and the reasons for the accident remain unknown. It was revealed that the propeller was missing, but not when or how it fell off.

In 1986, the novelist and former RAF pilot Wilbur Wright took up the challenge and asked the U.S. Air Force Information Center in California for the accident report on the missing Norseman. He was informed that no accident had been reported on that day and, in fact, no Norseman aircraft had been reported as missing throughout December 1944. Another mystery and another lie, as Wright subsequently discovered that eight Norsemen had been reported missing that month.

Wright then repeatedly wrote to every U.S. State Department and records office he could find, requesting information relating to the disappearance of Glenn Miller. But he was ignored until his letter of complaint to President Ronald Reagan encouraged a response out of the Military Reference Office. They confirmed there were several documents relating to the accident, but then failed to produce them. However, other departments continued to insist all records had been lost, destroyed, mislaid, or had never existed in the first place. When Wright telephoned George Chalou, the man in charge of the records office, to complain, he was alarmed by the latter's reaction during the conversation. According to Chalou (in a taped conversation with Wright): "They will never get them [the files] back either. Those files have been under lock and key for years and that is where they will be staying." There had been a cover-up after all.

After extensive research, Wilbur Wright's eventual conclusion was worthy of one of his own novels: that Glenn Miller probably had arrived in Paris the day before his band, where he was met by David Niven. Niven then set off to dramatically rescue Marlene Dietrich from the clutches of the Nazis, while Miller holed up in a brothel in the Parisian red-light district awaiting their return. Unfortunately, with time on his hands (and plenty of alcohol), he ended up becoming involved, and badly injured, in an unseemly bar brawl. The American authorities were horrified to discover the world's best-loved musician in a seedy brothel with a fractured skull. Miller was immediately airlifted back to Ohio, but he later died of his injuries.

Wright proposes three main strands of evidence. The first is based on the fact that David Niven makes no mention of Miller in his autobiography *The Moon's a Balloon,* published in 1971, despite the pair knowing each other well. Wright sees this as indicating Niven's awareness of the incident and his decision, for the sake of good grace and the Miller family honor, never to mention it again. (Indeed, he never even mentioned the name Glenn Miller to either his biographer, Sheridan Morley, or to his second wife.)

The second line of "proof" given by Wright is that Helen Miller soon moved to Pasadena, California, where she bought a burial plot with room for six graves. As her immediate family consisted of five people—herself and her son, daughter, and

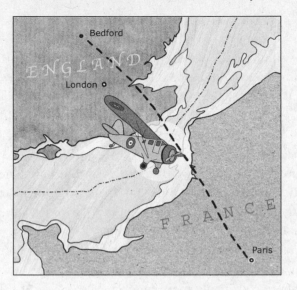

parents—it is therefore assumed that Miller himself occupies the last grave. When asked, the cemetery administrators denied Miller's presence but took a full fifteen months to reply to Wright's letter of inquiry, suggesting to Wright that both the family and local gravediggers were in on the cover-up. For him the clinching piece of evidence is that in 1954, a Parisian prostitute—still plying her trade opposite Fred's Bar, the brothel bar where Miller was alleged to have been drinking the night he went missing—revealed that her then boyfriend had told her what had happened to Glenn Miller, confirming the whole Parisian brothel story.

If that all seems a bit thin—and let's face it, it does—that's because the authorities only needed to remove one word and the whole cover-up would have been completely unnecessary. Think about the difference between reading "Glenn Miller died after being involved in a fight in a brothel bar" and "Glenn Miller died after being involved in a fight in a bar." That's it, no international outcry, just a respectable period of public mourning. No shame would have been heaped upon the Miller family and no extensive and complicated cover-up story would have been necessary. But if Wright's hypothesis is true, how could all those people who would need to have been involved for this story to have any basis in fact—including any witnesses, the French police, military personnel, flight crew, medics, doctors, nurses, administrators, gravediggers, family, friends, and probably Inspector Clouseau himself—have not failed to give the game away hundreds of times over the ensuing fifty years? Instead we have the silence of a film star, a six-berth burial plot, and the testimony of a Parisian tart well past her sell-by date.

My vote goes with the recent evidence that has emerged that Miller was on board the Norseman after all. The new story has a much more convincing explanation of the Americans' fear of the truth coming out. According to this theory, Miller boarded the Norseman at Twinwood Farm on December 14, 1944, just as Don Haynes said. The aircraft took off at 1:45 P.M. By 2:40 P.M. it was traveling through what was known as a jettison zone in the English Channel, an area set aside for returning bombers to drop their undischarged loads safely into the sea before they crossed the south coast. A fully laden bomber exploding on landing could wipe out an entire air base, so the jettison zone was stringently enforced. The only bomber to use the jettison zone that afternoon is known to have crossed it at around 3:40, at the time Miller should have been landing in Paris, and so it has never been thought relevant to the Miller mystery before. However, it has only recently been noticed that while the Miller flight would have been charted on Greenwich mean time (GMT), all military flight operations were logged using Central European time, which is one hour later. Therefore the bomber would have released its load directly over the area where Miller's Norseman would have been flying through, at a much lower altitude and in the opposite direction. Did the Americans hit their favorite musician with some not-so-friendly fire? There is certainly strong witness evidence to suggest they did, including some of the military aircrew themselves.

Fred Shaw, a navigator in one of the bombers, claimed, in an interview for an amateur film, that he saw the bombs his aircraft jettisoned strike a small plane beneath him. According to Shaw: "I had never seen a bombing before so I crawled from my navigator seat and put my head up into the observation blister. I saw a small high-wing monoplane, a Noorduyn Norseman, underneath us." Mr. Shaw claimed he didn't make any connection to the disappearance of Glenn Miller until he saw *The Glenn Miller Story* in 1956. "There is a kite down there, I told the rear gunner, there's a kite gone in," Shaw continued. "He then replied, yeah, I saw it too." At the time authorities had dismissed his claims as a publicity-seeking exercise, but Shaw remained adamant he had seen the small plane spiral out of control as a result of being hit.

In a sworn statement given on April 10, 1999, Fred W. Atkinson, Jr., a member of the 320th Air Transport Squadron responsible for taking Miller to Paris, stated the following:

You will recall in the movie, *The Glenn Miller Story,* the letter that Glenn Miller wrote to his wife that day [in which] he expressed the feeling that he might not see them again. Given the weather conditions and the type of aircraft that was a realistic probability.

Several days after our plane left London, we were notified that an aircraft that might be ours had crashed on the coast of France and that the occupants were dead. We dispatched a plane to that location and the aircraft and the bodies of our pilots were identified. Our crew also said that the other body definitely was that of Glenn Miller. They said there were identification papers and dogtags on his body. Our second crew that was in London at the time verified they had witnessed Glenn Miller and our two pilots board the aircraft and depart.

I recall the papers being processed to salvage our aircraft and report the death of our pilots on the squadron morning report. This report was turned in on a daily basis and notes the changes in status of all personnel as they occur. We had not experienced any deaths in our squadron until this time and this was a "double whammy" to us because of the loss of our pilots and the loss to the U.S. Armed Forces of probably the greatest morale booster (along with Bob Hope) that we all loved.

The flight logbook of another airman, Derek Thurman, appeared to corroborate the claim: "The bomb aimed down in the nose . . . saw an aircraft first, [and] remarked on it. The navigator shot out of his seat to have a look through a side blister [window] and he saw it sort of whip by, then the rear gunner said 'it's gone in,' sort of flipped over and went in. Whether it was brought down by a blast from one of the bombs, or was hit, is anybody's guess, really."

These three reports, all from independent sources, are consistent in the details they provide. The idea that a small aircraft could have been hit or damaged by an explosion nearby, thus causing its pilot to ditch it onto the beach, breaking its propeller, is not so far-fetched. And if so, the idea that the American military may have recovered the bodies, then dragged the prop-free plane back into the sea and created a cover story, is a racing certainty.

It tends to be the case that the first information to emerge from a suspicious incident such as the Miller mystery is the most accurate and reliable, especially where governments are concerned, as they won't have had time to concoct a story to suit their purposes. For my money, Miller was accidentally shot down by the very military he was traveling to Europe to entertain. The Miller family were told the truth, which explains the sixth burial plot, and in return for their patriotism in never speaking publicly of the accident, were handsomely compensated for their loss. David Niven, on the other hand, was warned he would never work in Hollywood again if he ever mentioned the matter to anybody, so he didn't; and the French prostitute was just looking to sell a story for enough francs to buy a new horsewhip and a couple of cheap bottles of Beaujolais.

It is hard to conceive of a more ludicrous story than the idea Glenn Miller was beaten up in a Parisian bordello and died of his injuries. In the case of Liam Gallagher, however, I doubt there would be any such cover-up if he was found dead in a Basra brothel. Although these days it's far more likely he would be stabbed in the school yard by a teenager after his mobile phone.

The Mystifying Death of a Media Mogul

....

Who finally stopped the bouncing Czech? The extraordinary life and death of Robert Maxwell.

In 1940, Jan Ludvik Hoch did what many young Jews in Eastern Europe were doing at that time, and ran away to England to fight the Nazis. The seventeen-year-old refugee then fought his way from the beaches of Normandy to the center of Berlin. After the war he went on to become a publisher, Labour MP, football club owner, company chairman, owner of the Mirror Group Newspapers, owner of the *New York Daily Times,* embezzler, and fraudster before finally slipping from the back of his yacht and into oblivion. The official autopsy report concluded the cause of death had been "accidental drowning," but, as in life, mystery shrouded the death of Jan Ludvik Hoch, a man who courted controversy from the moment he arrived in England and changed his name to Ian Robert Maxwell.

Maxwell joined the British army under a series of aliases, presumably because the War Office had suggested refugee soldiers should serve under invented names in case they were captured. Because he went by the names of Jones and du Maurier in addition to Maxwell, it is hard to find out much about what he got up to in the Second World War, although he did earn himself a medal. This was in January 1945 when his unit, the 6th Battalion of the North Staffordshire Regiment, was based at the river Meuse in Holland. He had recently been promoted to second lieutenant and his men were tasked with clearing a block of flats occupied by German soldiers. Maxwell led the assault and charged straight for the building, drawing heavy fire. Luckily for him, although not so luckily for his future employees, every bullet aimed at him missed. It was an act of bravery that won him the Military Cross.

But not all of his wartime exploits were quite as distinguished. His authorized biographer, Joe Haines, reveals how Maxwell's unit attempted to capture a German town by calling for the mayor to meet with Maxwell in a neutral location. He then told the mayor that the German soldiers would have to surrender or face destruction by mortar bombardment. In a letter to his wife, published in Haines's book *Maxwell* (1988), he wrote: "But as soon as we marched off a German tank opened fire on us. Luckily he missed so I shot the mayor and withdrew." Maxwell showed no remorse at killing an unarmed man in cold blood, and it was a sign of things to come.

As the war drew to a close, Robert Maxwell found himself working for the Control Commission, an Allied organization formed to manage the economy, state industry, and government of the defeated German people. His natural intelligence and gift for languages had been noticed by the high command of the Allied forces, and he soon found himself organizing various sections of the West German services, including the national newspapers. Back in Britain, his entrepreneurial spirit was quickly in evidence, and he became a shareholder in a London import and export company originally owned by a German, but Maxwell was soon in sole command.

Two years after the end of the war, Robert Maxwell's company was distributing scientific literature and manuals to both Britain and America after a deal was hatched with the German publishing heavyweight Springer Verlag (later Axel Springer) that established Maxwell in the marketplace. Another two years would

pass before he launched his own publishing company, Pergamon Press, after securing heavy investment from Springer. Such
was his initial success that he was able to buy Springer out of the
contract and take over as sole owner while settling in to the business of becoming seriously wealthy during the 1950s.

In 1959 he became the Labour candidate for Buckingham
and won the seat in 1964 as the new Labour government led by
Harold Wilson swept into power. He remained an MP until
1970, when the Conservatives under Ted Heath defeated Wilson
in the general election of that year. Maxwell too lost his seat but
by then had already realized that true power lay in journalism:
the pen really was mightier than the sword. In 1969 Maxwell had
unsuccessfully tried to buy the *News of the World,* having been
beaten to it by the Australian entrepreneur Rupert Murdoch.
Maxwell did not take defeat well and accused Murdoch of "employing the laws of the jungle," claiming he had made a "fair and
bona fide offer which has been frustrated and defeated over
three months of cynical maneuvering." In response Murdoch
stated that *News of the World* shareholders had judged him on his
record of newspaper ownership in Australia and were confident
in his ability. This was a clear slight on Maxwell's character as
well as the start of a bitter and lifelong rivalry between the two
men.

In 1969 Maxwell had opened negotiations with the American
businessman Saul Steinberg, who had declared an interest in
buying Pergamon Press Ltd. (PPL) on the understanding that
the company was making vast profits. Discovering this to be untrue, the American, despite months of negotiations, abruptly
pulled out of the proposed purchase. An investigation by the
Department of Trade and Industry (DTI) followed, in which in-
spectors revealed how transactions between private Maxwell
companies had been used to inflate the PPL share price. Steinberg initiated legal proceedings against the former MP, and in

1974 it was discreetly announced in *The New York Times* that he had received a payment of $6.25 million from Maxwell and his investment bankers. In their 1970 report, the DTI inspectors had concluded: "Notwithstanding Mr. Maxwell's acknowledged ability and energy, he is not in our opinion a person who can be relied upon to exercise proper stewardship of a publicly quoted company."

He lost control of Pergamon and the company's investment bankers appointed a new chairman in the shape of Sir Walter Coutts, who, with three independent directors, reversed the fortunes of Pergamon spectacularly and returned control of the company to Maxwell in 1974. Naturally, the "bouncing Czech," as he had become known because of his questionable integrity and his ability to bounce back from adversity, claimed the credit for the successful turnaround of his company. Coutts was later quoted by a biographer as saying: "Maxwell has an ability to sublimate anything that stops him getting what he wants. He is so flexible he is like a grasshopper. There is no question of morality or conscience. Maxwell is Number One and what Maxwell wants is the most important thing and to hell with anything else."

Building on the success of Pergamon, Maxwell bought Mirror Group Newspapers from Reed International for £113 million on June 13, 1984. Behind the scenes, he had already built up a mini empire consisting of, among other things, a record label, Nimbus Records, a printing company, a book publishing house, half of MTV Europe, 20 percent of Central Television in Britain, a cable television company, and two newspapers, *The People* and *Sporting Life*. As his empire, now called Maxwell Communications Corporation PLC (MCC), grew, so did his interest and influence in politics, especially as one of his various companies published the speeches of Chernenko, Ceauşescu, Brezhnev, Andropov, Kadar, Husak, and other Eastern Bloc leaders. He also published sycophantic biographies of world figures and used the opportunity to meet and interview them, which caused him to be ridiculed at home but strengthened his links with several totalitarian regimes.

Maxwell also claimed to have influence in Israel, and during a magazine interview for *Playboy* he boasted that it was he who had been responsible for persuading Israeli prime minister Yitzhak Shamir to exercise restraint in the face of Scud missile attacks from Saddam Hussein during the first Gulf War. He even boasted of close ties with the Israeli secret service, Mossad; indeed, after his death it was revealed that Maxwell had worked with the organization for many years and was known to a handful of elite Mossad agents by his code name "the Little Czech."

Journalists would later point out how Maxwell's companies would invariably take a downturn financially whenever Mossad was engaging in expensive covert operations, leading to speculation he was an important source of funds. Mossad was even rumored to have funded Maxwell's first big business venture, prompting suggestions that the whole of Maxwell's business empire was in fact a Mossad fund-raising venture. Stranger things have happened.

But in early 1991, the Little Czech was beginning to lose his bounce. A *Panorama* documentary for the BBC had drawn attention to the DTI's findings in the 1970s and suggested Maxwell had been bolstering the MCC share price through transactions with secretly owned companies in Liechtenstein and Gibraltar. The inevitable libel writs were issued, but a number of biographers followed with similar accusations. During the summer of 1991, Maxwell's relations with Israel soured when his repeated requests to Mossad to apply pressure on Israeli bankers to refinance his business were ignored. By now even the British Parliament was keeping a close eye on Maxwell's international business dealings.

Government ministers had known for a long time about Maxwell's influence with the various world leaders he had connections with. After all, it had been Maxwell who had liaised between Moscow and Tel Aviv during the August coup of 1991, in which the former head of the KGB Vladimir Kryuchkov and other hard-line Communists had attempted to oust Mikhail Gorbachev from power. (Who can forget the images being broadcast live by satellite during the early days of Sky Television of Boris Yeltsin standing on his tank with a bullhorn, organizing the defense of Moscow's White House?) Maxwell had been involved in arranging a meeting between the Israeli secret service and high-level KGB officials, including Kryuchkov, to discuss Mossad support for the plot to replace Gorbachev, the first Russian president to show any sign of being about to work closely with Western governments.

But then, according to the sworn testimony of former Mossad agent Ari Ben-Menashe, Maxwell had made the mistake of threatening Mossad with revealing information about the meet-

ing unless they supported him financially. According to Ben-
Menashe, both Maxwell and the *Daily Mirror*'s foreign editor
were longtime Mossad agents, and it was Maxwell who informed
the Israeli Embassy in London that Mordechai Vanunu had revealed details of Israel's nuclear capability to *The Sunday Times*.
Vanunu was immediately lured from his *Sunday Times*–provided
safe house in London to Rome, where he was snatched by
Mossad agents. He was later returned to Israel, convicted of trea-
son, and spent the next eighteen years protesting his innocence
from prison before finally being released in 2004.

On October 21, 1991, two members of Parliament, Labour's
George Galloway and the Tory Rupert Allison, were persuaded
to bring up the Vanunu affair, and Maxwell's part in it, in the
House of Commons. Protected by parliamentary privilege, in
which members could make allegations without fear of litiga-
tion, newspapers began to report a wide range of Maxwell-
related intrigues and mysteries.

The Israeli secret service was also concerned by the whole
Robert Maxwell situation. And as soon as he began to threaten
them, his fate was sealed. Agents quickly agreed to the meeting
of "great urgency" called for by a now cash-strapped Robert
Maxwell. They were well aware that Israel's reputation in the
West, particularly with America, on whom they were heavily re-
liant, would be severely tarnished should it become known that
they were in any way connected with the attempt to prevent
democracy in the Soviet Union. Mossad could no longer afford
to take any further risks with the Little Czech.

According to Ben-Menashe, in his book *Profits of War*, published in 1992, Maxwell was instructed to travel to Spain the following day where arrangements for a money transfer could be
made. His orders were to sail his yacht to Madeira and wait on it
there to receive further instructions. Maxwell breathed a sigh of
relief as he left England for Spain, believing his recent growing

financial problems were about to be solved. On October 31, he boarded the *Lady Ghislaine* at Gibraltar and berthed at Madeira, where he dined alone.

The following day a specialist Mossad team, in Spain to cover a series of Middle Eastern peace talks, was sent south. Maxwell then received a message instructing him to meet instead on the island of Tenerife on November 5. It is then alleged that Mossad agents boarded the boat near the Canary Islands during the night of November 3, removed Maxwell to another vessel, interrogated him throughout the following day, and then killed him by injecting air into his veins, which would have induced a heart attack.

But that's not the end of the story, because Ben-Menashe had also claimed that he personally delivered the CIA share of the profits from an earlier arms deal between Iran and Israel to Robert Maxwell in London and that Maxwell, in turn, was supposed to forward it on to America. Instead, the money disappeared into the gaping hole in Maxwell's balance sheet. Ben-Menashe alleged that Maxwell threatened the CIA with "damaging disclosures" should they press him for the return of the money. Now, I'm no expert in world politics, but when it comes to the wrong groups to annoy, the CIA and Mossad are two key ones to avoid, let alone blackmail.

It later transpired that as Robert Maxwell left for the Canary Islands he had also been told he was under investigation by the police for war crimes connected to the revelation about shooting that unarmed mayor in 1945. In March 2001, it was revealed under the Freedom of Information Act that weeks before he died detectives had started questioning members of Maxwell's former platoon but had yet to find any witnesses to the shooting. The former lieutenant was advised six months before his death that he faced a possible sentence of life imprisonment if found guilty. The Metropolitan Police file notes: "The reported circumstances of the shooting gave rise to an allegation of War Crimes. To some extent, the reporting of the shooting incident was confirmed by Mr. Maxwell in an interview he gave in 1988 to the journalist Brian Walden on 30th October 1988." Quite clearly there had been two shootings: by boasting of his wartime exploits, Maxwell had shot himself straight in the foot.

Following his death, Maxwell's body was released to the Israeli authorities, who performed a second autopsy, revealing that the injuries to the body were not consistent with falling off a yacht and that he "had probably been murdered." He was then afforded the honor of a Mount of Olives funeral, the resting place of Israel's most respected heroes. During the service, televised worldwide, the Israeli prime minister, Yitzhak Shamir, cryptically announced that Maxwell "had done more for Israel than can today be said." A remark that could simply be seen as a nod toward his fund-raising efforts for the Israeli secret service—or rather more.

Before the dust had, quite literally, settled on Maxwell's grave, calls of foul play could be heard. Maxwell's own daughter, Ghislaine, announced on television that his death had "not been an accident." Others insisted he had committed suicide to escape the shame of his collapsing empire, not to mention the jail sentence that would certainly have followed if he had been found guilty of fraud. But his life insurance company quickly paid out a thumping £20 million, indicating that they, at least, were certain Maxwell had not jumped. So, if his insurers had apparently ruled out suicide (and it was very much in their interests, after all, to prove that he had taken his own life), only two options are left to consider. Did Robert Maxwell have a heart attack and fall overboard, or was he murdered by the Israeli secret service? There is a further suggestion—put forward by those ever-busy conspiracy theorists—that Maxwell did not die but quietly slipped away from his problems, leaving another poor soul in the water to be found, misidentified, and then buried on the Mount of Olives.

Either way, his death was followed by a series of revelations about his controversial business methods, and accusations were made with impunity. It began to emerge that Maxwell had used £1 billion from his companies' pension funds in order to service his debt liability and fund his flamboyant lifestyle. MCC filed for bankruptcy protection in 1992 and Maxwell's two sons, Ian and Kevin, were declared the world's largest bankrupts, with debts in excess of £400 million. In 1995 they were charged with fraud but were acquitted in 1996. No doubt neither man was too impressed with his father's legacy.

Perhaps the final word should come from Lady Coutts, whose husband had rescued Maxwell all those years earlier. After dinner at Headington Hill Hall, Maxwell's country house and business headquarters near Oxford, the newspaperman was bidding his guests goodbye in some of the nine languages he, by then, boasted he could speak. When it came to Lady Coutts, she deliberately spoke to him in a language Maxwell had no knowledge of, Swahili: "*Kwaheri ashante sana sitaki kukuona tena,*" which means, "Goodbye and thank you very much. Now I never wish to see you again." And fortunately for her, she never did.

The Real Da Vinci Code:
The *Mona Lisa* Debate

....

*Who was the real Mona Lisa, who lurks
behind that famous smile, and where is her
portrait now? (Clue: it's not in the Louvre.)*

The debate about the identity of the model for the *Mona Lisa*,
with her enigmatic smile, has raged for more than five hundred
years. The bigger question, however, is not so much *who* is the
real Mona Lisa, but *where* she is.

Today Leonardo da Vinci (1452–1519) might have been a society portrait photographer, his job capturing images of the rich
and famous for a healthy fee. Back in the fifteenth century, this
work was both time-consuming and painstaking: it meant setting
up the easel and mixing the oils, and daubing away at just one
painting for many months at a time. For Leonardo it proved a lucrative day job, however, leaving him with enough time to concentrate on his other passions—engineering, sculpture, music,
to name but a few—and inventing tanks and helicopters five
hundred years before anyone else. And by the time he was in his
early forties, he was a successful portraitist, receiving commissions from a number of wealthy families.

Then, aged forty-eight, he started work on the painting for
which, besides *The Last Supper*, he is best remembered: a half-
length portrait of a woman in dark clothing set against an imaginary landscape. Unusually for portraits of the time, she gazes
directly out of the painting at the viewer, that famous smile play-

ing on her lips. There has been much speculation over the years as to the identity of Leonardo's muse. Giorgio Vasari, in his *Lives of the Artists* (1550), wrote that in 1499, Leonardo had been commissioned by Francesco di Zanobi del Giocondo, a wealthy merchant, to paint his new wife, twenty years his junior. Her name was Madonna Lisa del Giocondo, *madonna* being the equivalent in Italian for "madam" (or "my lady") and sometimes shortened to Mona—hence "Mona Lisa" (or "Madam Lisa").

Mona Lisa del Giocondo sat for Leonardo da Vinci between 1500 and 1505, and there is speculation that they became lovers. But this cannot be true, or at least it would have been unlikely, because—well, let me put it as delicately as I can—Leonardo was not that way inclined. He even once wrote: "The act of coitus and the members that serve it are so hideous that, if it were not for the beauty of faces, the human species would lose all its humanity." In 1505, Leonardo left Florence and presented Francesco with the portrait of his wife. The painting was unfinished, and Leonardo regarded it as a "work in progress," continuing to work on it whenever he returned to Florence.

A year before, in 1504, the artist Raphael is reported to have paid a visit to Leonardo's studio in Florence, where he made a sketch of the painting. This sketch depicts Mona Lisa sitting between two ornate Grecian columns, which do not appear in the painting as it can be seen today, hanging in the Louvre.

Mona Lisa del Giocondo was born in 1479, and would have been about twenty-one when Leonardo started to paint her. However, as most experts agree, the lady in the portrait appears to be older by at least ten years, perhaps even fifteen. So, what can have happened? Did Leonardo later paint out the columns and then make his subject appear rather older, or—extraordinary as this may sound—could the French have the wrong painting?

Important evidence can be found in the archives of Antonio de Beatis, secretary to Cardinal Luigi d'Aragona. In 1516,

Leonardo had been appointed "First Painter" to the new French king, Francis I. The cardinal visited the artist the following year and his secretary recorded their conversation. De Beatis specifically noted that the cardinal was shown three paintings: *Virgin and Child with St. Anne, St. John the Baptist,* and the "portrait of a certain lady from Florence, painted from life at the instance of the late Magnifico Giuliano de' Medici." Some researchers believe that Giuliano had fallen in love with Mona Lisa and had commissioned Leonardo to paint her for him. But this is a little difficult to believe when one considers how, these days, it would be hard enough trying to snap a digital image of another man's wife and keep it hidden from your own, let alone setting up with easel and oils and persuading the young lady in question to sit and be painted for months on end.

And, what's more, Giuliano—brother of Lorenzo the Magnificent and co-ruler of Florence with him—was murdered in the cathedral in Florence by the Pazzis, the rival family to the Medicis, in 1478. Mona Lisa del Giocondo, his purported lover, wasn't born until the following year.

Thus it appears that the portrait Leonardo took to France with him, and which remains there to this day, is not of our Mona after all. The subject of the painting is now believed to be Giuliano's real lover, Costanza d'Avalos, a happy-go-lucky girl with such a light and friendly nature that she became known as "the smiling one"—or *la gioconda* in Italian—and that is how the painting transported to France is better known throughout continental Europe.

The similarity between Costanza's nickname and the surname del Giocondo, of the real Mona Lisa, is an obvious source of the confusion at the heart of the mystery. Then, it was discovered that in Giovanni Paolo Lomazzo's book on painting, architecture, and sculpture (his *Trattato dell'arte della pittura, scultura ed architettura,* published in 1584), the historian refers to both the *Mona Lisa* and *La Gioconda,* suggesting that Leonardo did not change his painting by removing the pillars and making Mona Lisa look older, because there were in fact two pictures. Of these only one found its way into the hands of the French, and that one is neither of Mona Lisa nor called *Mona Lisa.* Lomazzo dedicated his book to the Grand Duke of Savoy, a known expert on the work of Leonardo da Vinci, making it highly unlikely he was mistaken.

More recently, an inventory taken in 1520 by Giacomo, one of Leonardo's disciples, has been discovered, listing the "work" (the *Mona Lisa,* by implication) as still being in Leonardo's personal collection at the time of his death (perhaps returned to him by Francesco for him to finish?), and subsequently recording that the painting in the hands of the French royal family at Fontainebleau was called *La Gioconda* ("The Smiling One"). Which confirms that the painting Napoléon later had in his bedchamber in the Tuileries in 1804, and which was subsequently transferred to the Louvre in 1805, is not the *Mona Lisa* but *La Gioconda.*

So what became of the real *Mona Lisa,* complete with Grecian pillars and younger subject? It turns out that the painting remained in Italy for over two centuries until it was bought by an English nobleman, who hung the work at his manor house in Somerset without realizing the importance or significance of the painting. It is possible he didn't even know that the portrait was by Leonardo da Vinci, as the artist is unlikely to have signed the unfinished picture. Then Hugh Blaker, a London art dealer, discovered the piece in 1914 and was able to buy it for only a few guineas. He took the painting back to his studio in Isleworth, London, and thereafter the painting became known throughout the art world, rather unromantically, as the Isleworth Mona Lisa, which, no doubt, would have pleased the Florentine Mona Lisa no end.

Now that the real *Mona Lisa* had been identified, the final piece of evidence revolves around the question of which of the pictures is finished. Vasari, in his *Lives of the Artists,* states that "Leonardo worked on the *Mona Lisa* for four years and then left it unfinished," which is correct, but also that "the painting is now in the possession of Francis, King of France, at Fontainebleau." Here he was mistaken: the portrait in the Louvre certainly looks as if Leonardo had completed it, unless he intended to cheer her up a little at some point in the future. But then Vasari gives a description of the unfinished *Mona Lisa:*

> The eyes had the luster and watery sheen that is always seen in real life and, around them, were those touches of red and the lashes that cannot be represented without the greatest of subtlety. The nose, with its beautiful nostrils, rosy and tender, seemed to be alive. The opening of the mouth, united by the red of the lips to the flesh tones, seemed not to be colored but to be living flesh.

Now, I'm no art expert, but that doesn't sound like the picture in Paris to me. Is her mouth open, for instance? And there are also no eyelashes or eyebrows on the figure in the Louvre painting. Giorgio Vasari was born on July 30, 1511, and he was only six years old when Leonardo took his paintings to Paris, meaning he must have been describing the portrait given to Francesco del Giocondo in 1505 (and possibly returned to the artist later for finishing), and not the portrait in Paris. And as the art dealer Hugh Blaker later confirmed, this description perfectly matches the unfinished painting he bought in Bath. Other experts clearly agreed as, in 1962, a Swiss syndicate led by the art collector Dr. Henry F. Pulitzer bought the painting for millions of pounds.

So it appears that there are, after all, two enigmatically "Smiling Ones" in the world: the Costanza d'Avalos *La Gioconda* hanging in the Louvre in Paris that has bedecked more tea towels, calendars, and biscuit tins than any other painting in the world, and the wonderful, unfinished Mona Lisa *Gioconda* that used to hang in Isleworth, London, and is now safely stored in the vault of a Swiss bank as part of the Pulitzer collection, treated with the respect she deserves. After all, she was a respectable Florentine lady married to a popular merchant and not the tawdry mistress of a gangster murdered in a turf war five hundred years ago. Still, we must allow the French their tourist attraction, especially as they cut the heads off their others some time ago.

If Gentlemen Prefer Blondes, Who Killed Marilyn Monroe?

....

*Did the celebrated sex symbol take her own life,
or did someone lend a hand?*

The fourth of August 1962 seemed an ordinary Saturday in the life of Marilyn Monroe, insofar as her life could be called "ordinary" at this stage, the world-famous actress now beset by depression and paranoia. Despite daily therapy with her personal psychiatrist, Dr. Ralph Greenson, who lived nearby, Marilyn's anxiety attacks and bouts of depression had worsened. She had accidentally overdosed, and her stomach had been pumped, on more than one occasion.

Marilyn had become increasingly dependent on Dr. Greenson, and she consulted him constantly about her troubled love life, which, by this time, had included relationships with two Kennedy brothers, Frank Sinatra, the baseball star Joe DiMaggio, the playwright Arthur Miller, and the scientist Albert Einstein. She also believed that both the Mafia and the FBI, not to mention the CIA, were keeping an eye on her in the wake of the Profumo scandal that summer, in which Russian spies had compromised English cabinet minister John Profumo by fixing him up with a young prostitute. And she was right to worry, because Monroe's relationships with both JFK and Bobby Kennedy—right in the middle of America's crisis over the Soviet plan to base nuclear missiles in Cuba, within striking distance of the mainland—had led to her being considered a serious security risk.

Marilyn had spent the previous evening at home, and in good spirits, with her press agent and best friend, Pat Newcomb, who had then stayed over. But when Pat arose the following day, Marilyn appeared "grouchy" and claimed not to have slept very well. Her housekeeper, Eunice Murray, later called in Dr. Greenson after Monroe asked her if there was "any oxygen in the house." As the afternoon progressed, Marilyn's condition deteriorated: she appeared increasingly drugged and lethargic. Greenson had been trying to break Monroe's dependency on Nembutal, but knew she had received a new prescription the previous day. He knew, furthermore, that supplies of her favorite barbiturate were stashed around the house and that she could have taken these at any time.

Pat Newcomb left at 6 P.M. After another session of therapy, Greenson left at 7 P.M. At 7:15 Joe DiMaggio, Jr., her ex-husband's son, dropped by; Marilyn was happy to learn he was breaking off his engagement to a woman she did not like, and DiMaggio, Jr., later confirmed the actress was in high spirits by the time he left, as did Dr. Greenson, whom she had rung shortly afterward to inform him of the good news.

Then Eunice claims to have awoken at 3 A.M. to see a light shining under Marilyn's bedroom door and a telephone cable leading from a socket in the hallway into the bedroom, both of which were highly unusual. Finding the door was locked, the housekeeper telephoned Dr. Greenson, who rushed over, broke into the bedroom via a window, and, at 3:50 A.M., found the Hollywood actress lying naked, facedown and clearly dead. However, the veracity of their account began to seem more questionable when it emerged later in the investigation that not only would Monroe's deep-pile bedroom carpet have ensured that no light could have escaped from the room, but the door had no working lock. The plot appeared to thicken further when it was revealed that Arthur Jacobs, Monroe's publicist, had been informed of her death at between 10 and 10:30 P.M. the previous evening: he could confirm the time as he had to leave a musical performance of another client to arrange the "press issues." So we know that before poor Marilyn's body was even cold, a tissue of lies had already started to be spun.

The autopsy, carried out by Dr. Thomas Noguchi—who was later to conduct the high-profile autopsies on Natalie Wood and Robert Kennedy—concluded that Marilyn had died as a result of acute barbiturate poisoning. This led the psychiatric experts involved with the inquest to a conclusion of "probable suicide." But Los Angeles County prosecutor John W. Miner—who had attended the autopsy and who was privy to all the facts surrounding her mysterious death—was furious. He didn't believe then that Monroe had taken her own life, either deliberately or by accident, and today, over forty years later, he still doesn't. So what really did happen to the celebrated Hollywood actress?

Norma Jeane Mortenson arrived in the world at 9:30 A.M. on June 1, 1926, at Los Angeles County Hospital. Her mother, Gladys Pearl Monroe Baker, had already walked out on Norma's father (well, her father according to the birth certificate at least), ostensibly because he had become "boring." Gladys was later diagnosed with hereditary paranoid schizophrenia, a mental condition that also afflicted her mother and father and which had contributed to the deaths of two of her grandparents.

When Norma Jeane was only seven years old, her mother was committed to a "rest home" and the little girl was then moved around to various foster parents and institutions. With nowhere to live—her latest foster parents were moving to the East Coast and couldn't take her with them—she got married, to James Dougherty, just two weeks after her sixteenth birthday, in June 1942.

Since America was now at war, her new husband joined the navy and Norma Jeane went out to work. At just seventeen, she was already drinking heavily and suffering from depression. As a little girl, she had dreamed of stardom: "Even as a child I used to think as I looked out on the Hollywood night that there must be thousands of little girls sitting alone like me, dreaming of becoming a movie star. But, I thought, I'm not going to worry about them. I'm dreaming the hardest." It must have seemed a faraway dream when she was clocking in at the munitions factory every morning at 7 A.M.

During the summer of 1944, *Yank* magazine commissioned a feature on young American women at work for the war effort. Private David Conover had been moving along the assembly line taking pictures of the most attractive employees when he came upon a young blonde who was busy fitting propellers. Although her face was covered in dirt and grease, he stopped in his tracks, stunned by her unusual beauty. Private Conover immediately offered Norma Jeane five dollars an hour to model for him, and the resulting pictures attracted the attention of the Blue Book modeling agency. Within a year, Norma Jeane had been featured on the front cover of no fewer than thirty-three national magazines, catapulting the young lady toward national stardom. Her first marriage proved an early casualty of her obsessive, meteoric rise.

In July 1946, one month after her twentieth birthday, Norma Jeane secured a contract with Twentieth Century–Fox. The studio wanted her to have a more glamorous name and, after a few duff suggestions, the casting director, Ben Lyon, came up with "Marilyn," after his own favorite actress, Marilyn Miller. Then Norma Jeane offered her mother's maiden name. The studio director wearily asked what it was, but his eyes lit up when she replied, "Monroe."

There then followed four years of success and failure in both her career and love life, leading to, after the sudden death of a lover, her first real suicide attempt, when she swallowed a bottle of sleeping pills. Throughout the 1950s, Marilyn became more and more ubiquitous, appearing in hundreds of films, TV shows, musicals, and radio broadcasts. By the end of the decade, Norma Jean had become Hollywood's golden girl, mixing with the rich, famous, and powerful. But the recognition she craved didn't make her happy. Her marriages to two much older men, each highly acclaimed in his field, clearly illustrated her search both for security and for a father figure. Her short union with Joe DiMaggio was quickly followed by her third marriage, this time to America's most celebrated playwright, Arthur Miller. It was after her divorce from Miller, in 1961, that things began to go badly wrong. Was it really a coincidence that this was when her affair started with the most powerful man in the world, President John F. Kennedy?

After the divorce from Miller, Marilyn, increasingly dependent on alcohol, barbiturates, and Dr. Greenson, became friends with the English actor Peter Lawford and his wife, Patricia, a sister of JFK. It was at one of their parties that she first met the Kennedy brothers. Unsurprisingly, this drew the attention of the FBI, whose head, J. Edgar Hoover (see also "John Dillinger: Whatever Happened to America's Robin Hood?" page 77), was obsessed with building a file on the growing sexual adventures of the president and his brother Robert, the attorney general. The Mafia were also taking a close interest in the actress. The FBI among others believed that the Kennedy brothers' father, Joe, had been a partner of the infamous Mafia don Frank Costello during the Prohibition years. It was said that decades later, when JFK ran for president, the old man had called on the Cosa Nostra to help buy votes. Some Mafia members believed the Kennedys then owed them a favor or two and expected a close, lucrative relationship with the Kennedy administration once John had taken office.

So they were furious when Bobby Kennedy, the newly appointed attorney general, made it his personal crusade to crack down on organized crime, making the wrong sort of enemies in the process, many of whom vowed revenge. Even so, most Mafia members realized the Kennedy family, the biggest mob of them all, now had public opinion firmly on their side, not to mention all the state police forces and the U.S. military at their instant beck and call. Any act of revenge on the Kennedys would have to be carefully thought out, more carefully than the customary sort of Mob hit on a rival family member. In 1962, exposing the Kennedys' many infidelities to the press was thought the best tactic to diminish public support for the brothers. Monroe had found herself in bed, so to speak, with some of the most dangerous people in the world, and still didn't realize it. Instead she was naïvely dreaming of becoming America's First Lady.

Marilyn's love affair with the president became common knowledge among the American power set during the first six months of 1962, but remained unknown to the public. Hoover's FBI was busily building a file detailing Monroe's movements and had even, some believed, placed listening devices inside her home. Increasingly worried by her "chattering" about their relationship, the president was even more alarmed by his brother-in-law's discovery that she kept a detailed diary of their sexual encounters and what they had discussed. JFK abruptly ended the affair in July, using his brother Bobby as the messenger. Unfortunately for the administration, Bobby too then fell under the actress's spell. Marilyn, still bitter from her rejection by the president, did not reciprocate his feelings, but she embarked on a love affair with him nevertheless.

Marilyn had no intention of marrying the smitten younger Kennedy, however, even on one occasion asking Dr. Greenson, "Oh, what am I to do about Bobby?" Greenson was more concerned about the psychological damage such affairs were doing to his client and about her personal safety. The international threat to America was from the Cuban missile crisis and the domestic problem was coming from the Mafia. Marilyn knew too much about too many people, mobsters and politicians alike, and more than one group was worried that she might spill the beans. Her increasingly erratic behavior had turned her from a trophy blonde to an outright liability. When Bobby unceremoniously broke off their affair by having the private telephone line he had installed for her disconnected, Marilyn was devastated. She bombarded the White House switchboard with telephone calls but was never connected with either Kennedy. Distraught, she had told friends—including Peter Lawford, JFK's brother-in-law—that she planned to "come clean" about her relationships with both brothers in revenge for the way she felt she had been treated by the pair.

However, in July 1962, during the final two weeks of Marilyn's life, there were reports that she was feeling more positive about the future than she had been. She had received several new offers of film parts, her friends were many and supportive, and, despite everything, she was still optimistic about reviving her relationship with the president.

In this frame of mind, she happily accepted an invitation from Frank Sinatra to a weekend at the Cal Neva Resort on Lake Tahoe, believing the Kennedys to be behind the invitation. Accounts of this weekend differ, but they are all highly colored. One goes that Monroe was taken aback to discover the brutal gangster Sam Giancana was there, apparently to warn her against creating problems for the brothers. Another version has Joe DiMaggio arriving unexpectedly at the lodge and becoming furious with both Sinatra and the Kennedys for luring his ex-wife there, plying her with drugs and alcohol, and taking compromising photographs to be used as blackmail should she ever threaten to expose her affairs with John and Bobby. The following weekend, Marilyn was found dead at her home in Brent-

wood, California, having apparently committed suicide, the subsequent mystery and intrigue surrounding her death involving some of the best-loved and most influential people on the planet.

According to the official version of events, after Joe DiMaggio, Jr., left at around 7:30 P.M., Peter Lawford then phoned Marilyn at 7:45 P.M. to invite her to a party. He testified Monroe sounded heavily drugged—somewhat contrary to the upbeat mood reported by DiMaggio—and that she failed to respond several times before shouting her own name repeatedly into the phone. Lawford then quoted how Marilyn had ended the conversation: "Say goodbye to Pat, say goodbye to the president and say goodbye to yourself because you are a nice guy." She then hung up.

The next official evidence we have is the statement of Eunice Murray, the housekeeper, who claimed to have seen the light on under the bedroom door at 3 A.M. and telephoned Dr. Greenson. He then confirmed he arrived and broke in through Marilyn's bedroom window at 3:50 A.M. to find the actress dead, at which point he telephoned the police. At 4:25 A.M. Sergeant Jack Clemmons of the Los Angeles Police Department received a phone call from Dr. Engelberg, Marilyn's personal physician, who told him his patient had committed suicide. Given what we know about the evidence today, it would have been quite impossible for Engelberg to diagnose suicide at that stage, although Clemmons is adamant that that is what he was told.

When the police officer arrived at the scene, he noted three people with the body, Eunice Murray, Dr. Greenson, and Dr. Engelberg, who led Clemmons to it and made a point of bringing to his attention the bottles of drugs on the bedside table. Clemmons noted: "She was lying facedown in what I call the soldier's position. Her face was in a pillow, her arms were by her side, her right arm was slightly bent. Her legs were stretched out perfectly straight." The policeman's immediate reaction was that she had been placed in that position. Having been at the scene of numerous suicides, he knew that, contrary to what most people believed, victims of an overdose of sleeping tablets tend to suffer convulsions and vomiting before they die, often ending up in a contorted or twisted pose.

The testimony of the three witnesses convinced Sergeant Clemmons that they were lying. Publicly all three witnesses maintained their original story that the body was found at 3:50 A.M.; privately they stated the body had been discovered four hours earlier but they had been "not allowed" to contact the police until Twentieth Century–Fox had given them permission. Clemmons then noted that no light—let alone the telephone cable reported by Eunice Murray—was able to pass under the bedroom door and that it had no working lock. Crucially, there was no drinking glass in the room, or indeed any kind of receptacle that could have contained the water or alcohol Marilyn would have needed to swallow so many pills.

The police officer took a closer look at the window Dr. Greenson claimed to have broken to gain access to the room, and found broken glass on the outside, consistent with the window having been broken from inside the room and not from the outside.

The autopsy conclusions were that, judging by the high level of sedatives—eight milligrams of hydrate and four milligrams of Nembutal in her blood count and a much higher concentration, thirteen milligrams of Nembutal, in her liver—and the absence of any signs of foul play, Marilyn had taken her own life. These findings were soon disputed by some key forensic experts, however, who pointed out that no traces of Nembutal had been found in either her stomach or intestinal tract. There was also no evidence of the yellow Nembutal capsules, which would not have fully dissolved by the time the autopsy took place. An injection was ruled out because no needle marks were found and because such a high dose would have caused instant death plus residual bruising around the site of the puncture mark. So, as Marilyn appeared to have taken nothing orally and nothing directly into the veins, forensic experts concluded that the drug had been administered by way of an enema. This was consistent with the bruising on the victim's lower back and would account for the "abnormal discoloration of the colon." In other words, the drugs that killed her must have been introduced anally.

Now, I'm no expert, but I think that most people would agree that to prepare a fatal cocktail of drugs and then push it up your own backside is an unlikely way to commit suicide. So despite Monroe's famously erratic behavior and ongoing depression, suicide has been ruled out by every mental-health professional reviewing her case. Indeed, it is alleged that detailed notes made of taped conversations with her psychiatrist only a week before her death reveal her as anything but suicidal. Yet those tapes, along with other vital evidence and statements, have all gone missing. John W. Miner has been consistently clear in his views: "Marilyn Monroe bears the stigma of suicide. That is wrong and must be corrected."

So, if we are to rule out suicide, then there are only two other possibilities for us to consider: accident or murder. But if Marilyn did die as a result of rectally administered barbiturates, then it is hard to see how that could be an accident. Let's be honest, who could claim that they pushed a poisonous drug up Marilyn Monroe's rear by accident—and surely she would have noticed?

Marilyn's psychiatrist, Dr. Greenson, and her physician, Dr. Engelberg, were working together to reduce the insomniac actress's Nembutal dependency by replacing it with chloral hydrate, but taken together they are a powerful and dangerous mixture. One suggestion is that Engelberg had given Monroe a further prescription of Nembutal and forgotten to inform Greenson. As Engelberg was having serious marital problems at the time, other, more personal matters may have occupied his mind.

Perhaps Marilyn—who once commented, "Yes, I enjoy enemas, so what"—had been taking Nembutal throughout the day, explaining its presence in her liver and blood. Without knowing this, Dr. Greenson could then have prepared a chloral hydrate enema to be administered by Eunice Murray, which became deadly on interaction with the Nembutal. Any doctor would be reluctant to admit to such a mistake, especially in relation to such a high-profile patient, and this would perhaps explain the many discrepancies in the stories of those who found the body and the apparent staging of the scene that the police were unhappy about. It would also explain evidence that the body was discovered at 10 P.M. and not 3:50 A.M., along with an ambulance driver's account that Marilyn was taken to hospital in a coma before midnight where she died before the body was returned and "found." Eunice Murray would certainly wish to stay quiet, as it would have been she who administered the fatal dose. So this is quite a powerful theory. But if it is true then the doctors involved would be guilty only of negligence, and certainly not murder.

But if not an accident, could it have been murder? Several witnesses have placed Bobby Kennedy at the scene on the day of the death. There is also reliable evidence that he removed Monroe's diaries and other notebooks. Her angry telephone calls to the White House and the fear of her speaking publicly were a real problem for the Kennedy administration. Indeed, it would be a mistake to rule out certain governments at certain times taking drastic action to prevent a scandal. But as Bobby Kennedy approached most matters from a very high moral standpoint, it is hard to believe he would personally hold Monroe down and push barbiturates up her bottom to kill her. He made no secret of his visit to Monroe that day and was seen there by many people, including a policeman. Are we to seriously believe he—or somebody he was with on that day—killed Monroe? After all, Marilyn Monroe, the drunken actress who famously sang a very breathy "Happy Birthday, Mr. President" at Madison Square Garden in front of the world's media, hardly posed a major threat to the most powerful government in the world. What classified information could she alone possibly have had that no one else had access to? The promiscuity of the Kennedy brothers was well known in 1962. Might the government's fear of the headline "I Slept with the President," by Marilyn Monroe, have led to her assassination, when the chorus of replies would probably have been: "Who hasn't?"

The idea that the American government silenced Marilyn sounds like a classic urban legend, the kind of story that just grows over the years, so that the more times it is told the more true it appears to become. We know Marilyn died of a drug overdose and we know it could not have been suicide. So why push a killer drug up her bottom when a staged car accident, drowning, bullet in the head, or even a drugged drink that would leave residue in the stomach to make her death look like suicide would have been a better option? If the Kennedy government intended to kill off everybody who had embarrassing information or who opposed their administration, then why should they start with a scatty film star? And as for the idea that Marilyn Monroe was murdered by the Mafia, who staged it to look as if the Kennedys had been involved, that seems even less likely when a cold, clear look at the evidence is taken. Nobody can explain why anybody trying to make a murder look like suicide would use a killer enema—it just doesn't make sense. If the Mafia— who had perfected, by then, their concrete boots technique— had really wanted to get rid of her, then Marilyn is more likely to have vanished without a trace, either swimming with the fishes or wrapped inside a freeway overpass.

Instead, I think the biggest clue lies in a comment alleged to have been made on the night of Monroe's death by Dr. Greenson: "God damn it, he has given her a prescription I didn't know about." So it seems after all that the death of the most famous woman on the planet might well have been a simple, tragic accident caused by the people Marilyn most relied on.

The Piano Man

••

Who was the strange castaway found in Kent,
apparently unable to communicate
but a brilliant pianist?

On April 7, 2005—in a case very reminiscent of the story of Kaspar Hauser (see page 117)—the police were called to deal with a stranger wandering the streets of Sheerness in Kent, England. The man was respectably dressed in a suit and tie, but soaked to the skin. As it hadn't been raining, it was assumed he had been washed ashore and was disoriented and frightened after his experiences at sea. He appeared unable to speak and carried no form of identification. Police officers took him to the Medway Maritime Hospital and placed him in the care of social workers. Although the man seemed to be in reasonable shape physically, he still showed no signs of being able to understand anything or communicate in any way. Instead he simply sat and stared around the psychiatric ward he had been placed in.

Both national and international publicity failed to reveal his identity, and staff at the hospital continually tried to communicate with the man. Finally, after being handed pencil and paper, the enigmatic stranger, instead of jotting down his name or any other useful details, sketched a grand piano. The story goes that he was then led to a piano, where he gave a note-perfect virtuoso performance of a Tchaikovsky piano concerto to the astonishment of hospital staff. In the words of one of his carers, the social worker Michael Camp: "He seems to come alive when playing the piano, for several hours at a time."

At this point, the world's media began to take a keen interest in the stranger, now dubbed the "Piano Man," and hundreds of leads were followed up after people claimed they recognized him. First he was believed to be a Czech concert pianist or a well-known Canadian eccentric. He was then thought to be a French street musician or a German genius. Other theories suggested his voice box had been removed, or that he was mentally ill or possibly autistic.

Before long, Hollywood was taking an interest in the story of the "autistic piano genius." But the *Mirror* soon tired of the story and began dismissing the man as a hoaxer who could in reality barely play a single note on the piano and was actually conversing freely with medical staff. And that is where the mystery deepens, because rather than issue a statement to the contrary, providing recorded evidence—which anyone could have done, just using their mobile phone—the NHS flatly refused to comment. I can smell those rats again . . .

But it would be another few months before the Piano Man made the news again, having now left England, it would appear, as mysteriously as he had arrived. News came on August 22, 2005, in the shape of an announcement by the National Health Trust:

> The patient dubbed the Piano Man is no longer in the care of the West Kent NHS and Social Care Trust. He has been discharged from our care following a marked improvement in his condition. The rules regarding patient confidentiality mean that the Trust is unable to make any further comment on this story. This includes any comment on his condition, current location or the circumstances in which he left the Trust's care.

And that was that, he was gone. Naturally newshounds scurried around for a story, but all they could come up with was a statement from the German Embassy, where a spokesman said: "The hospital called us up on Friday morning saying that they had a man there claiming to be a German national. We contacted his parents and his identification was confirmed. We then gave him replacement travel documents and he left the UK using his own arrangements on Saturday morning."

So had it all just been a hoax? Had he, as one national newspaper claimed, simply given the game away and revealed all to the staff at the hospital? But if he was a simple con man, what was the con? Walking around soaking wet in Sheerness and then staying silent in a psychiatric ward for four months is hardly a con, is it? And as his name and place of origin have never been revealed, he never really benefited from the publicity. And could he play the piano or couldn't he? Even I can tell the difference between a performance of Tchaikovsky's piano concerto and a rendition of "Chopsticks" by a one-finger plinker-plonker—and, thanks to Pete Townshend, I'm nearly deaf. No, the real mystery here is who at the hospital was telling tales about virtuoso piano recitals and why. And also, how did the stranger manage to leave England without the British press getting hold of at least his identity, because that doesn't happen very often, does it? A story as unusual as this simply fading out and disappearing—I don't think so.

However, despite no official announcement of his identity ever being made, there are claims that the Piano Man has been identified—as a former newspaper columnist and mental health care worker from Germany—while his lawyer has apparently issued a statement explaining that his client may have been experiencing a "psychotic episode." Further investigation reveals many other claims attributed to unnamed sources. Meanwhile a couple, claiming to be his parents, are said to have insisted their son had told them he "had no idea what happened to him. He suddenly woke up one day and remembered who he was." It all sounds a bit fishy to me, so I am off to try and find out right now and will let you know if I discover anything.

The Dreadful Demise of Edgar Allan Poe

....

The unexplained death of the master of gothic horror

It was Election Day in Baltimore, Maryland. Ryan's Tavern, a popular saloon bar, had doubled up for the day as a polling station, and men had been shuffling in and out to cast their votes since daybreak. Many stopped for some light refreshment before going about their business, but few of them took any notice of the resident drunks slumped in the corners, propped against tables, or generally scattered around the bar. Then, for reasons that are unclear, a voter named Joseph Walker went over to help one of them. The man, in a state of confused desperation, called out the names of people he appeared to know until finally Walker recognized one and immediately sent a note to Dr. Joseph Snodgrass, which read: "There is a gentleman, rather worse for wear, at Ryan's Fourth Ward Polls, and who appears to be in great distress. He says he is acquainted with you and I assure you he is in need of immediate assistance."

Just five days later, on October 8, 1849, the Baltimore *Sun* published a somber notice:

> We regret to learn that Edgar Allan Poe, Esq., the distinguished American poet, scholar and critic, died in this city yesterday morning, after an illness of four or five days. This announcement, coming so sudden and unexpected, will cause poignant

regret among all who admire genius, and have sympathy for
the frailties all too often attending it.

Yet Poe wasn't supposed to have been in Baltimore at all; he was
meant to have been in Philadelphia for a business meeting, followed by a journey to New York to meet his former mother-in-law,
Maria Clemm. Edgar Allan Poe never arrived in Philadelphia,
and Maria Clemm was never to see him again.

The dark events and insecurities of his life were dramatized
throughout Poe's writings, and it's possible that his mysterious
death was connected with someone very close to him. Edgar Poe
was the son of traveling actors. He was not yet four years old
when his parents died, within a few days of each other, and the
three Poe orphans (Edgar had an elder brother and a younger
sister) were separated and sent to live with different foster families in Richmond, Virginia. Edgar was taken in by John and
Frances Allan, a wealthy, childless couple who raised him as their
own. As a sign of respect for his foster parents, Poe later adopted
their surname as his middle name and thereafter became
known by the name for which he would become famous the
world over: Edgar Allan Poe.

But a serious rift developed between Poe and his foster father
when Edgar returned from college in 1827 with large gambling
debts that John Allan angrily refused to pay. Shortly afterward
Poe joined the army, achieving the rank of sergeant major before returning, in 1829, for the funeral of his beloved foster
mother, Frances. The following year John remarried, and when
the new Mrs. Allan promptly produced three sons, she became
openly hostile to the grown-up foster son she had inherited.

This reached crisis point in March 1834 when Poe discovered
that John Allan was gravely ill. He rushed to his bedside, only to
find the route blocked by the second Mrs. Allan. When Poe angrily pushed past her, he was confronted by a furious John Allan,

who cursed him from his deathbed, banishing him from the house. Poe then discovered, after Allan's death, that the man whom he had once lovingly called "Pa," and whose affections he had relied upon as a small boy, had changed his will, removing any mention of him.

While Poe was at college, he began writing poetry, anonymously publishing his first collection, *Tamerlane and Other Poems*, in 1827. In 1831, he turned his attention to the short stories of mystery and the macabre that he was to become famous for. They were instantly popular. Before long, Edgar had progressed from mere contributor to editor at the *Southern Literary Messenger*.

Throughout all this, his ties to his real family remained very strong, and they became stronger when in 1836, aged twenty-seven, he fell in love with his thirteen-year-old cousin Virginia Clemm. Despite Virginia's being so young, the two married within the year, with the full blessing of his aunt (and mother-in-law) Maria Clemm, who then became the third mother figure in the young writer's life.

In 1839, he accepted the job of both editor and contributor at *Burton's Gentleman's Magazine* in Philadelphia and, during his time there, wrote the macabre tales "William Wilson" and "The Fall of the House of Usher." It was the popularity of psychological thrillers like these that saw his personal reputation flourish, and in 1841 Poe had completed his most enduring tale, "The Murders in the Rue Morgue," featuring, for the first time, his fictional detective C. Auguste Dupin. The story was truly unique in the sense that it introduced a new and popular genre in which a series of seemingly unconnected clues are presented to the reader and not drawn together until the final scene, in which the murderer is unmasked in front of the other characters by the detective. The style had never before been used in literature, and Poe's sleuth is credited with being the first fictional detective in the history of storytelling, paving the way for Arthur Conan Doyle's Sherlock Holmes and Agatha Christie's Hercule Poirot.

However, it was Edgar's poem "The Raven," published in 1845, that signaled his true rise to fame, with the public queuing up for Poe's lectures just to hear the writer perform his work in person. The effect in 1845 was something like a modern song-writer or musician would achieve with a number-one hit single. Other successful poems followed, and Poe's popularity contin-ued to increase until disaster struck in 1847, when his beloved wife, Virginia, died. Edgar was heartbroken, and his grief is be-lieved to have inspired the short poem "Deep in the earth my love is lying / And I must weep alone." Her death was to signal the beginning of Poe's downhill struggle leading to his own mys-terious death only two years later—a period that was marked by alcoholism, depression, a suicide attempt, and several failed ro-mances. All of which was accompanied by a desperate attempt to raise funds to support his beloved mother-in-law and for the launch of his own publication, *The Stylus*. (Despite his literary success, much of his own money had been spent on drink.)

Then, during the summer months of 1849, things started to look up again. Poe, who was once again out on the lecture cir-cuit, met Elmira Shelton, an old childhood sweetheart, back in Richmond and they rekindled their romance. With Elmira's en-couragement, Poe joined the Sons of Temperance movement and renounced alcohol. He wrote to Maria Clemm: "I think she loves me more devotedly than any one I ever knew & I cannot help loving her in return. <u>If possible</u> I will get married before I start—but there is no telling."

And it wasn't just his love life that had turned the corner. His lecture tour was also proving to be a great success and he had gathered over three hundred annual subscriptions for his pro-posed new magazine, at five dollars per year. This would mean

Poe was in funds to the tune of at least $1,500, a considerable amount in 1849. He was due to leave Richmond for his next engagement in Philadelphia, where he had been commissioned by a wealthy piano manufacturer, John Loud, to spend two days editing his wife's collection of poems. The fee was to be $100, a large sum for two days' work, and Poe had eagerly accepted the commission. He then intended to leave Philadelphia and continue to New York. Here he would collect Maria Clemm and her possessions and bring her back to Richmond, where he intended to settle down with Elmira.

Before leaving Richmond on September 27, Edgar visited his physician, Dr. John F. Carter, and, after a short conversation, walked to the Saddler's restaurant on the opposite side of the road, absentmindedly taking Carter's malacca cane instead of his own. There he met acquaintances, who later walked with him to catch the overnight boat to Baltimore from where he would catch the train to Philadelphia. They left him "sober and cheerful," promising to be back in Richmond soon.

Poe had written to Maria Clemm advising her that "on Tuesday I start for Phila. to attend to Mrs. Loud's poems—& possibly on Thursday I may start for N. York." He also asked her somewhat cryptically to write to him at the Philadelphia post office, addressing the letter to E.S.T. Grey, Esq., and suggested that rather than turning up at her house, he should send for her instead on his arrival in the city. It is not clear why he needed to use a false name in Philadelphia or why he felt unable to visit the house in New York. Was he in debt, perhaps, or in some kind of danger?

Nothing more is known for sure about Edgar Allan Poe's movements until he turned up disheveled and disoriented at Ryan's Tavern in Baltimore five days later, on October 3. Apart from his failing to keep his appointment in Philadelphia with Mrs. Loud, that is. And there are various theories why he didn't.

One account claims he fell ill as soon as he arrived in Philadelphia and, intending to catch another train to New York, boarded at the wrong platform and returned to Baltimore by mistake. A second account makes the same claim, but suggests that he was drunk rather than sick.

When a guard on the train to Philadelphia claimed he had witnessed Poe being "followed through the carriages" by two mysterious men, speculation arose that friends of Elmira Shelton, possibly her brothers, had followed the writer, suspecting he was having a liaison with another woman, and then had forced the writer back to Baltimore, beaten him into a stupor, and left him on the street, where he wandered into the bar and was discovered. Meanwhile another theory suggests that Poe had been in regular correspondence with a lady with whom he subsequently quarreled. When Edgar refused to give back her letters, she sent the men to enforce their return and they then beat up her former lover. Were they the two men on the train— assuming the guard's testimony is to be believed and there were any mysterious men in the first place?

Lending substance to this last claim is the suggestion that prior to meeting Elmira again, Poe had been engaged to a wealthy widow after only a brief courtship in what some regarded as a callous attempt by the writer to gain funding for his new magazine. This was broken off after a violent confrontation between a drunken Poe and his terrified fiancée, and it is possible that this lady had been the sender of the letters Poe had refused to return. In addition, rather than just being simple love letters, they may have contained a promise of funding that Poe intended to later claim as a contractual obligation. Hence the rather extreme measures the lady had to resort to in order to get them back.

Though varied and unreliable, each account is consistent with the idea that Poe did not stay in Philadelphia and possibly did not even leave Baltimore in the first place. He certainly failed to collect the letter from Mrs. Clemm addressed to E.S.T. Grey, because the post office, as was common practice, published receipt of it in the Philadelphia Public Ledger on October 3, 1849, the same day that he lay dying in the bar in Baltimore. Such was Poe's devotion to Maria Clemm, it seems unlikely he would not have made straight for the post office to collect a letter he was expecting if he *had* arrived in Philadelphia as planned.

But while there were no confirmed sightings of Poe in Baltimore during the week prior to his death, the writer's physical condition offers some clues as to what may have happened. Writing to Maria Clemm on November 15, 1849, Dr. Moran (the doctor at the hospital to which Poe was admitted) noted:

> Presuming you are already aware of the malady of which Mr. Poe died I need only state concisely the particulars of his circumstances from his entrance until his decease.
>
> When brought to the Hospital he was unconscious of his condition—who brought him or with whom he had been associating. He remained in this condition from 5 Ocl. in the afternoon—the hour of his admission—until 3 next morning. This was on the 3rd October.
>
> To this state succeeded tremor of the limbs, and at first a busy, but not violent or active delirium—constant talking—and vacant converse with spectral and imaginary objects on the walls. His face was pale and his whole person drenched in perspiration. We were unable to induce tranquility before the second day after his admission.
>
> Having left orders with the nurses to that effect, I was summoned to his bedside so soon as conscious supervened, and questioned him in reference to his family—place of residence—relatives etc. But his answers were incoherent & unsatisfactory. He told me, however, he has a Wife in Rich-

mond (which I have since learned was not the fact) that he did not know when he left that city or what has become of his trunk of clothing.

The most obvious clue lies in a reference to his clothing. Dr. Snodgrass later described what Poe had been wearing at the time he was found:

> His hat—or rather the hat of somebody else, for he had evidently been robbed of his clothing, or cheated in an exchange, was a cheap palm-leaf one, without a band, and soiled. His coat was of commonest alpaca [cheap camel fleece], and evidently "second hand"; his pants of gray-cassimere [plain wool], dingy and badly fitting . . . his shirt was badly crumpled and soiled. On his feet were boots of coarse material, giving no sign of having been blackened for a long time, if at all.

Edgar Allan Poe had not been so attired when he left Richmond, so this is the first real evidence of foul play. Snodgrass wondered if Poe had not fallen off the wagon with a vengeance and sold or exchanged his own clothes for more liquor. A weeklong drinking binge could well have had fatal consequences, but

as Snodgrass didn't know at the time, Poe had been in possession of a considerable sum of money when he arrived in Baltimore, which was now missing, and even he would have had
trouble drinking through $1,500 worth of whiskey in a week.
This theory of the demon drink reclaiming Poe has often been
repeated over the years, but it is worth remembering that the
main architect of such an idea is Dr. Snodgrass himself, who
later became famous during the 1850s for his temperance lectures and often used the famous writer as an example of what
can happen should a person succumb to the evils of alcohol.
With the benefit of hindsight, it is easy to look back at the now
collected evidence and see that Snodgrass was not averse to a little exaggeration. For example, in his written account, "The Facts
of Poe's Death and Burial," published in May 1867 in *Beadle's
Monthly,* he transcribed the note he had first received from
Joseph Walker. Where Walker describes Poe as "rather worse for
wear," Snodgrass changed the wording to "in a state of beastly in-
toxication."

Dr. Moran also made a career out of his deceased patient by
lecturing and writing about Poe for many years. He flamboyantly
claimed in his *Defense of Poe,* published in 1885, that Edgar Allan
Poe's final words to him had been: "He who arched the heavens
and upholds the universe has His decrees legibly written upon the
frontlet of every human being and upon demons incarnate."
While they certainly sound like the words of a delirious man, I am
unable to decide who was the more so—Poe or the good Dr.
Moran. And this is completely undercut by Moran's letter to
Maria Clemm on November 15, 1849, in which he reports that
the writer's final offering was "Lord, help my poor soul," which
would mean the only reliable information from Dr. Moran was his
initial description of Poe when he was brought to the hospital.

In his first letter to Maria Clemm, Dr. Moran refers to Poe's
trunk, which was discovered at a hotel a few days later. But

Moran fails to mention Poe still had the key in his pocket, despite having apparently had his clothing stolen, or that he still had Dr. Carter's malacca cane, which Moran sent to Maria to be returned to the doctor in Richmond. And why would the thief hand Poe back the key to his trunk having presumably forced him to change clothes?

Even more suspiciously, there has been no mention of the large sum of money Poe was known to have had in his possession on his arrival in Baltimore. Some are puzzled by the revelation that Poe's trunk was booked in to a hotel, when the writer was only supposed to be passing through on his way to Philadelphia, but September 28, 1849, was a Friday, and in his letter to Mrs. Clemm, Poe informed her he was traveling on Tuesday, October 2, so it is quite conceivable Poe had checked in to a hotel with the intention of meeting somebody in Baltimore for the weekend, although as yet nobody has been able to ascertain who that was.

Four days after Poe's death, on October 11, 1849, his cousin Neilson Poe wrote to Mrs. Clemm, claiming to have carried out an exhaustive inquiry as to Poe's movements during that final week, but with no success: "Where he spent the time he was here, or under what circumstances, I have been unable to ascertain." However, within a few weeks, Neilson had written to Poe's first biographer suggesting he had acquired some information about Poe's death, which was "known only unto me."

This curious remark needs investigating. Because if Poe died at the hands of another, as the evidence tends to suggest, then how could Neilson be the only one who knew anything about it, unless he himself was involved? At the time, Neilson promised to write it all down in a "deliberate communication," but nothing was ever sent to Poe's biographer and Neilson is not known to have ever written anything about the death of his famous cousin.

It is interesting to note that this was the same cousin who raced to the bedside of Poe when he was taken ill, only to be refused entrance by Dr. Moran, claiming Poe was too delirious to receive visitors; and yet Neilson claimed only a week later to Maria Clemm that he had no knowledge of Edgar's presence in Baltimore. The question is, was he lying, and was it Neilson whom Edgar had planned to meet over that weekend in Baltimore? Moreover, did Dr. Moran suspect Neilson had something to do with Poe's condition and is that why he refused him access?

Election Day, the day Poe was discovered, was a dangerous time during the mid-nineteenth century, as a practice known as "cooping" was widely employed by unscrupulous politicians and their supporters in many American cities. William Baird explained these goings-on in a paper published in Baltimore during the mid-1870s:

> At that time, and for years before and after, there was an infamous custom in this and other cities, at election time, of "cooping" voters. That is, gangs of men picked up, or even carried off by force, men whom they found in the streets and transported them to cellars in various slums of the city, where they were kept under guard, threatened, maltreated if they attempted to escape, often robbed, and always compelled to drink whiskey, sometimes mixed with other drugs, until they were stupefied and helpless.
>
> At the election these miserable wretches were brought up to the polls in carts or omnibuses, under guard, and made to vote the tickets in their hands, repeatedly at different voting places. Death from the ill treatment was not uncommon. The general belief here is that Poe was seized by one of these gangs then "cooped," stupefied with liquor, dragged out and voted again and again, then turned adrift to die.

The cooping theory is one suggested by most Edgar Allan Poe biographies and accounts of his death. In those days, Baltimore elections were notorious for corruption and violence, with political parties willing to resort to extreme measures to ensure the success of their favorite candidates. Poe was discovered on Election Day after he had been missing for five days, and he was found lying in an apparently drunken stupor in a bar where the votes were actually being cast.

But as with all the other theories about Poe's death, the cooping hypothesis has an obvious flaw. Edgar was well known in Baltimore and therefore likely to be recognized by many people. Cooping being a dangerous and highly illegal activity, it is unlikely "coopers" would risk holding Edgar Allan Poe with others who would be able to identify him. The Whigs were the major political party, headed by Zachary Taylor, who had been elected president in 1848, and it turns out that a delegate of the Eighteenth Ward had been none other than Poe's cousin, Neilson Poe.

Could Neilson have been involved in cooping his famous cousin? This would certainly explain why Poe had checked in to the hotel for the weekend, on the assumption that he would be spending an exciting few days with his politically active cousin in the run-up to an important election. Could Poe have been drugged by Neilson, who had then stolen his money? This would account for Poe's incoherent state at the time he was found. But perhaps Neilson had no intention of killing Edgar. Maybe he assumed instead that Poe would regain normal consciousness in the bar, unaware of what had happened to him. Which would explain Neilson's panicked dash to the hospital when he found out that Poe had been admitted, despite claiming later he could find no trace of Edgar in Baltimore all week. Or perhaps it had been slightly more innocent and Neilson might have rescued Poe from the coopers for his own political party but, wishing to keep the matter a secret from the electoral authorities, simply left him to recover and be found. He might have slipped the trunk key into Edgar's pocket knowing he had a change of clothes in town and could tidy himself up when he had recovered.

In a letter of November 27, 1874, N. H. Morrison claimed to J. H. Ingram (another of Poe's biographers): "The story of Poe's death has never been told. Neilson Poe has all the facts but I am afraid may not be willing to share them. I do not see why. The actual facts are less discreditable than the common reports published. Poe came to the city in the midst of an election and that election was the cause of his death." What Neilson really knew about the death of his cousin has never been fully established, but it is clear that he knew something.

Poe's remains are interred next to those of his grandfather General David Poe, Sr., the American Revolutionary War hero, in Baltimore cemetery, and every year, on the anniversary of Poe's birthday, on January 19, fans still assemble for a silent vigil. Every year since 1949 a smartly dressed hooded man leaning on a silver cane has approached Poe's grave, knelt in respect, toasted the writer with a glass of cognac, and left the bottle, along with three red roses, at the graveside. Poe enthusiasts have watched this ritual without ever attempting to identify the stranger. In 1993, a note was also left, stating that "the torch will now be passed," and since then a younger, similarly dressed man has carried out the ritual.

Over time the debate about Poe's death has served only to make his mystery seem more mysterious and increase the intrigue. But the tragic event remains one of the most mysterious deaths in literary history, and one cannot help concluding that the great man might have been secretly pleased about that. He might also have a few questions to ask of his cousin Neilson.

It's Raining Frogs

....

Next time it clouds over, consider taking a fishing net with you instead of your umbrella.

On Wednesday 9th February 1859 I was fetching a piece of timber when I was startled by something falling all over my head and down my neck. On reaching down my neck I was surprised to find they were little fish. By this time the whole ground was covered in them and when I took off my hat the brim was full of fish, all jumping about. They covered the ground in a long strip of about eighty yards by twelve yards that we measured. My mates and I collected many buckets full with our hands.

That was the sworn testimony of John Lewis of Aberdare in Mid Glamorgan, Wales. Since then there have been thousands of reports from all over the world of showers of either small fish or tiny frogs. These so-called fish or frog "falls" have never been properly explained, however, and remain one of nature's mysteries.

Moreover, the phenomenon goes back farther than the nineteenth century. In 1666, a letter to a fellow of the Royal Society went largely unreported, perhaps because it was also the year of the Great Fire of London:

On the Wednesday before Easter, Anno Domini 1666, a Pasture Field at Cranstead near Wrotham in Kent, about two

acres, which is far from any part of the sea or river and a place where there are no fish ponds and a scarcity of water, was all overspread with little fishes, conceived to be raining down. There having been at the time a great tempest of thunder and rain. The fishes were about the length of a man's little finger and judged by all that saw them to be young Whitings. Many of them were taken up and shown to several other people.

But even farther back in history, in the second century B.C., we come across the following account in the *Histories* of Heraclides Lembus:

In Paeonia and Dardania, it rained frogs and so great was their number that they filled the houses and streets. Well, during the first days the people killed them and shut up their houses and made the best of it. But before long they could do nothing about it. Their vessels were filled with frogs, which were found boiled or baked with their food. Besides, they could not use the water, nor could they step upon the ground because of the heaps of frogs piled up and they also became overcome with [such] disgust at the smell of the creatures that they fled the country.

A whirlwind or waterspout picking up the contents of a lake and depositing it elsewhere has been put forward as a possible explanation for the phenomenon, but nobody really knows why fish or frog falls occur. Nor can anybody explain to me why it's raining cats and dogs outside my window today.

The Terrifying Affair of
Spring-heeled Jack

*Was the devil really out and about
in nineteenth-century London?*

In January 1838, the lord mayor of London, Sir John Cowan, was opening his morning mail when one letter in particular grabbed his attention. It was from a citizen living in Peckham who had been terrified by a demonic figure while crossing Peckham Rye. The mysterious beast had also pounced on a young lady and left her too frightened to give even a vague description, apart from the fact that it appeared to bounce, rather than walk normally like a man. The letter read:

TO THE RIGHT HONOURABLE LORD MAYOR.

My Lord—The writer presumes that your lordship will kindly overlook the liberty he has taken in addressing a few lines on a subject which within the last few weeks caused much alarming sensation in the villages within three or four miles of south London.

It appears that some individuals (from, as the writer believes, the higher ranks of life) have laid a wager with a mischievous and foolhardy companion (name as yet unknown), that he dares not take on himself the task of visiting many of the villages near London in three disguises, a ghost, a bear and a devil. And moreover that he dare not enter gentlemen's gardens for the purpose of alarming the inmates of the house.

The wager has however been accepted and the unmanly villain has succeeded in depriving seven ladies of their senses . . .

The affair has now been going on for some time, and, strange to say, the papers are still silent on the subject. The writer is very unwilling to be unjust to any man, but he has reason to believe that they have the history at their finger-ends but, through interested motives, are induced to remain silent. It is, however, high time that such detestable nuisance should be put a stop to.

I remain your Lordship's most humble servant,

A RESIDENT OF PECKHAM.

The letter forced the mayor to go public with his concerns and, as soon as he did, reports flooded in from all parts of the capital, earning the strange creature official recognition by the authorities. The newspapers soon nicknamed the beast "Spring-heeled Jack" and published stories of appearances on a regular basis. Hundreds of reported assaults were recorded—it seemed that the mysterious beast was terrorizing several districts of the city—and it wasn't long before many Londoners were refusing to leave their houses after dark.

A young maid, Mary Stevens, was intimidated on Barnes Common. Polly Adams, a barmaid in south London, was attacked as she walked across Blackheath. A servant girl at the home of Mr. Ashworth in Whitechapel answered the door to a hooded beast with fiery breath, and her screams were heard hundreds of yards away. Lucy Scales, a teenager from Limehouse, reported an assault. And Jane Allsop was virtually strangled by the hooded creature, until her family fought it off.

To local press hounds, the young women all described a "hideous face and his eyes were balls of fire. He had icc claws and breathed fire." With a photofit picture like this, you would think Spring-heeled Jack shouldn't have been too hard to find, especially if you were standing next to him in the omnibus

queue. It was a description echoed by many Londoners, who also reported seeing him bounce across the rooftops to make his escape.

That old English warhorse the Duke of Wellington, by then nearly seventy, armed himself to the teeth and went in search of the creature, but without success. Reports of attacks continued to flood in, and the army set a trap. After nightfall, sentries reported the beast suddenly appearing in front of them and "slapping their faces with icy hands" before bouncing away. Reports continued to arrive, mainly from the outskirts of London, in the areas of Ealing, Chiswick, Hammersmith, and Kensington.

Rumors abounded of servant girls dying of fright as soon as Jack appeared and of children being found ripped to shreds in the street—although there is no evidence of any of this and it is almost certainly untrue, so you can sleep easy. By the end of the century, Spring-heeled Jack had spread his wings to Surrey, Devon, and Hampshire and made his last appearance in Liverpool in 1904. Presumably he had his springs nicked and was forced to retire.

The true identity (or identities) of Spring-heeled Jack has never been discovered, although, despite frightening people and giving them the odd slap around the face, he had committed no serious crimes. History is littered with pranksters, and my guess is that "Jack" was one, or several, of these.

Beware of USOs

··

(That's Unidentified Submersible Objects to you.)

If there are intelligent beings on planets so far away that we haven't discovered them yet and they are watching us, then their technology must be far more advanced than ours. So forget those saucer-shaped tin cans people kept photographing during the 1950s and '60s, as they clearly couldn't have traveled that kind of distance. Even so, it does lead you to wonder where minds immeasurably greater than our own would hide an observation post to keep a closer eye on us human beings. It occurred to me that since the first lunar landing (assuming you believe Neil Armstrong walked on the moon in 1969), we have gained a pretty good understanding of everything within 240,000 miles of our planet, and obviously, thanks to the Hubble Space Telescope, way beyond, but we still know very little about things that are right under our noses.

Consider the oceans, for example. The farthest-reaching submarine was the remote-controlled Japanese *Kaiko* surveillance sub, an unmanned craft designed for deep-sea observation. The *Kaiko* could reach depths of nearly thirty-eight thousand feet, which sounds impressive initially until you work out that that comes to a grand total of just over seven miles. So we can see, thanks to the Hubble telescope, a distance of between thirteen and fourteen billion light-years upward, but only seven miles downward. Now, we already know that unless you are Bigfoot (see page 25) or the Loch Ness Monster (see page 121) there is nowhere on earth to escape the long reach of satellites or modern radar systems, so if I were an alien capable of traveling thirteen billion light-years to come and spy on us, I would set up home underwater, happy in the knowledge that nobody would find me. Better still, directly underneath one of the polar ice caps.

So that led me to research unidentified submersible objects, the deep-sea version of UFOs. My attention was quickly drawn to a small port on the southern coastline of Nova Scotia in eastern Canada. But I'd like to point out that this had absolutely nothing to do with its name. Shag Harbor is normally a very quiet place, but on October 4, 1967, it suddenly became a hive of activity. At 11:20 P.M. a group of eleven people watched as a low-flying object suddenly veered downward at an angle of forty-five degrees and plunged into the water. Some reported a bright flash of light as it hit the surface, while others claimed they saw four or five glowing orange lights. Laurie Wickens, a Shag Harbor native, jumped onto the harbor wall to get a better view and said he saw the UFO floating on the surface with a strange orange light glowing on top of it.

Believing it was an airplane crash, residents immediately called in the Canadian Mounted Police, who were then joined by the U.S. military, suspiciously quickly. Within half an hour, local fishermen had put together a civilian rescue team and were already at the scene of the accident. But they were puzzled to see no signs of any debris, wreckage, oil, or bodies—only a large patch of foaming yellow bubbles. When one of the men attempted to take a sample by dipping his net into the water, the bubbles failed to attach themselves and the net always came up clean.

By the following afternoon, the authorities were satisfied that no aircraft had been reported missing and the area was sealed off while divers combed it for clues. The official report of the incident revealed nothing, although it was later leaked that a second, identical craft had soon joined the first under the water and that after a short delay, they both rose to the surface and zoomed away. Thirty years later, one of the navy divers, interviewed for a television documentary, claimed the U.S. military had monitored the two USOs for several days before losing contact with them. To this day nobody knows what happened at Shag Harbor, although the sheer number of witness statements—all consistent in timing and in their descriptions of the size, color, and speed of the craft, coupled with the evidence of the yellow foam observed by most of the initial rescue team— would appear to provide credible evidence of a kind of underwater USO activity that had never been seriously considered before.

And the events at Shag Harbor are by no means the only sightings. During the prolonged Cold War (which followed the very hot Second World War), many submarine commanders reported tracking mystery underwater vessels, often in Norway and other Scandinavian countries, that, when cornered in one of the fjords, would mysteriously vanish. On September 4, 1957, *The Daily Telegraph* reported that three uniformed police officers had witnessed a red, circular USO emerging from the depths of the Bristol Channel and taking a westerly route toward Wales. In the Lake District many sightings of USOs have been reported since the 1980s, and on one occasion during 1994 twenty-two people reported observing two underwater craft at Derwent Water for five minutes before these disappeared without a trace. The Lakes now receive as many as one sighting every year, the latest being in December 2004, leading to suggestions that beings from outer space have set up observation posts beneath the tranquil waters.

One sighting in 1977 was independently confirmed by no fewer than ten policemen. Soon after midnight on August 28,

officers claim to have witnessed a large diamond- or triangular-shaped object close to Lake Windermere, the largest natural lake in England. PC David Wild was the first to spot the strange craft, and he watched it hover above the A592 at fifteen hundred feet for twenty-five minutes before it vanished in front of his very eyes. Two other officers also witnessed the same phenomenon some distance away, and John Platt described seeing what looked like a "massive seagoing catamaran with two hulls," adding that the "surface was a dull charcoal color and giant lights were mounted on the front." Which sounds exactly like a catamaran with headlights to me. But despite Windermere being such a vast body of water, it is only 220 feet deep—hardly the best place in the world to hide a colony of aliens.

More recently, the Dutch submarine *Bruinvis* (yes, the Dutch do have submarines) reported an underwater collision with a "solid object" on October 19, 2001, in the Sognefjord in Norway. Most of the crew clearly heard the noise and the sub limped back in to port for emergency repairs. Navy divers later confirmed damage to the underside of the vessel. But it was in August of the previous year, in the Barents Sea at the northern tip of Norway, that one of the world's worst submarine disasters took place, the sinking of the Russian vessel the *Kursk*. Could it have been involved in a collision with a USO?

The *Kursk*, the flagship submarine of the Russian Northern Fleet, was proudly launched in 1995. But less than six years later, the world held its breath when the Russian authorities announced that an accident had caused the submarine—with 118 men on board—to sink to the bottom of the ocean. One team of rescuers reported that major damage to the front section had rendered the escape hatch useless, but that there were also deep gashes along the side to the fin at the rear, suggesting the cause of the accident had not been an explosion, as was first thought, but a collision with an unidentified object. Yet neither the U.S. nor the Royal Navy, which also had submarines in the area, were able to report a collision with any of their own craft.

Furthermore, the *Kursk*'s periscope and external masts were fully extended, suggesting that the submarine had been operating within ten meters of the surface when it was struck, as these sections of the vessel are always fully retracted, even during emergency dives, in deeper water. Who can forget the rescue team's harrowing reports of the hammering made by the surviving sailors as they tried in vain to save them? The subsequent salvage operation revealed that at least twenty-three men had remained alive for many days in the dark and cold, hoping for a rescue that never materialized. When the craft was eventually brought to the surface, it was revealed that a neat circular hole had been punched into the side, unlike the damage made by a torpedo or collision with another submarine, and the front section was almost completely torn away.

To this day, nobody knows what collided with the *Kursk*, although the rescue teams later described some green and white marker buoys bobbing on the surface that then mysteriously disappeared. (Such buoys are used for alerting passing vessels or aircraft that an accident has occurred; Russian vessels use only red and white rescue buoys, however.) Russian sources later confirmed that when the *Kursk* was eventually located there was a second, large object lying next to it on the seabed, which slowly moved away and then disappeared altogether. For weeks afterward, Russian attack submarines and warships of the Northern Fleet closely guarded the entire area. But whatever it was that sank the *Kursk,* with the loss of 118 lives, remains unknown to this day.

And it's not just in far northern waters. The Japanese deep-sea submersible the *Kaiko* mysteriously disappeared in May 2003 in the Pacific Ocean close to Japan after the steel cable attaching it to the mother ship, the *Kairei,* inexplicably snapped, and it has never been seen since, leading to all kinds of speculation about mysterious forces lurking deep in our oceans.

But I've got the same problem with USOs that I have with all UFOs. If there are life forms from other galaxies that have found planet Earth and discovered it inhabited by, in most cases, intelligent life, then why don't they just land, shake hands, and introduce themselves? If we were to discover life on Venus, for example, would we buzz around their planet scaring the crap out of everybody living there? I doubt it very much; I expect we would do what the great adventurers of the past centuries have done when they discovered new lands. And that is to introduce themselves politely to the natives and then steal all their diamonds and other mineral wealth. They wouldn't lurk around for decades first, would they? They'd wade straight in. So instead of laboring too hard over this problem, I think it may be time to pay a visit to Shag Harbor. Not to investigate any ongoing USO activities, but to find out how it came by its interesting name.

The St. Valentine's Day Massacre

....

*Who was really behind the notorious mass shooting
in Prohibition-era Chicago?*

On the evening of February 14, 1929, Chicago police made a grisly discovery. Inside a garage complex at 2212 North Clark Street lay the bodies of seven well-dressed men, who had all been brutally executed.

The investigators were puzzled. The victims were all mobsters with violent reputations who worked for the Irish bootlegger George "Bugs" Moran. As Moran's gang was known to be feuding with other gangsters, they should have been heavily armed and fully prepared for one of the shootouts that were becoming increasingly common in Prohibition-era Chicago. How had so many of them ended up unarmed in a run-down warehouse in the backstreets of the city? And why had none of them fought back—indeed, how could such experienced criminals have been led so tamely to their fate? It was a mystery to the police and a mystery to Bugs Moran. The American press and public wanted to know what could have possibly led to the horrific events of that bleak winter's night.

The place to start in any murder investigation is motive: finding out who would benefit most from the killing. The motive in this instance was obvious, and the person likely to benefit most from the killing seemed pretty obvious too. It was the height of the Prohibition era and many mobs and gangs were competing for the lucrative (and illegal) trade in alcohol, drugs, gambling, and prostitution. Bugs Moran had formed an impressive smuggling and supply racket in Chicago. He also had a small army of followers, mainly from the Irish community. Taking on the Irishman would be akin to going to war, which ruled out all the small-time operators. For a suspect, the investigators kept returning to one name and one name alone, Al "Scarface" Capone.

Capone's gang of Italian mobsters were well known to the authorities. His network of prostitutes, gambling dens, smugglers, bootlegging, and protection rackets had created an impressive empire, and he was estimated to be worth in the region of $65 million, a staggering sum of money in 1929, equivalent to approximately $7.2 billion today. He was a force to be reckoned with in Chicago and his policy of expansion through killing his business rivals placed him atop the list of suspects. It seemed obvious that he was behind it. But he denied all knowledge. His rival, Moran, had been neither killed nor even threatened, and the men lying dead in that garage were mere foot soldiers whose death could not have benefited Capone in any way. He had also been in Florida on Valentine's Day.

Even the single eyewitness to the shooting couldn't shed any light on the identity of the perpetrators. The police had found Frank "Tight Lips" Gusenberg lying amid the carnage and choking on his own blood. He was rushed to the hospital and, on finally regaining consciousness, was immediately asked who had shot him. Gusenberg carefully looked around the room before replying, "Shot? Nobody shot me!" He died soon afterward and the general belief was that he had recognized somebody in the room, although his silence hadn't helped him survive.

The police returned to the scene and tried to piece together the events leading up to the shooting from what little evidence they had. It was statements from the inhabitants of North Clark Street that provided their first real breakthrough. Several residents confirmed they had heard gunfire but swore they had then seen two uniformed policemen leading two civilians away at gunpoint. The two "suspects" had been handcuffed and bundled into a police car and then driven away. Reassured that the police were already present and everything appeared to be under control, no one made any effort to report the matter to the authorities. However, the Chicago police had no record of any shootings or arrests made on North Clark Street on the night of February 14. The investigators followed up every clue and lead they had, but they were all dead ends and no convictions were ever secured for the brutal murders in the warehouse on that cold February night.

Even though there was no proof linking Capone to the massacre, Bugs Moran had gotten the message. He promptly moved his gang out of the North Side, leaving all business in that area for the Italians. But he had already made a major error by commenting publicly to a journalist, "Only Capone kills like that." These five short words were a serious breach of the gangster code of silence, after which even his own gang members began to lose respect for their boss.

Moran became an increasingly marginalized and desperate figure. In 1946, he was finally arrested for robbing a bank messenger of $10,000, a far cry from the high-level crime and luxurious lifestyle he had formerly enjoyed. Moran was sentenced to ten years' imprisonment but was immediately rearrested on his release. He was given another ten years at Leavenworth Federal Penitentiary, where he died of cancer in 1957. His body lies in a pauper's grave within the prison walls.

The St. Valentine's Day Massacre also led to the downfall of Al Capone himself, because it brought his activities to the attention of the federal government. Despite no evidence being found to connect him to the killings on North Clark Street, the gangster

was soon convicted on charges of income tax evasion and, in 1931, sentenced to eleven years at the notorious high-security prison at Alcatraz.

While in prison, Capone's mental health began to deteriorate: toward the end he was convinced that the ghost of James Clark, one of the St. Valentine's Day victims, was haunting him. It was the only clue he ever gave of any involvement in the killings. After his release, Capone spent the last seven years of his life quietly at his luxury estate near Palm Beach, Florida. On January 25, 1947, he died of a heart attack thought to have been caused by the third-stage complications of syphilis.

Meanwhile the garage on North Clark Street—the site of the infamous events—was demolished; the area is now a landscaped car park for a nursing home. The infamous wall Moran's men were shot against was dismantled, sold at auction, and shipped to Canada, where it was rebuilt in the toilets of a Vancouver theme bar, the Banjo Palace. When that business closed down, each brick of the famous wall was sold off, as macabre souvenirs.

The St. Valentine's Day massacre itself remained a mystery

until recently. The true events of that fateful night were discovered long after the deaths of everybody involved. In January 1929, Jack "Machine Gun" McGurn, one of the Capone mob, was making a telephone call on the street when Peter and Frank Gusenberg's car drew alongside. When the two Moran mobsters recognized McGurn, they opened fire, but missed him, which was to prove a major error for the brothers. Capone and Bugs Moran were struggling for control of the bootlegging business in Chicago, and the tension between them had begun to degenerate into street warfare. But with many other mobsters muscling in on the action, it was sometimes unclear who was responsible for which act of violence. This time there was no mistake. McGurn knew exactly who had tried to kill him.

Capone was already aware of the might of Moran's army and a month or so earlier had secretly discussed with an associate how to eliminate the "Moran risk." When he was allegedly warned he would "have to kill a lot of people to get to Bugs Moran," Capone joked that he would send plenty of flowers. So when "Machine Gun" McGurn approached his boss with a plan to avenge the phone-booth shooting, Capone saw the perfect opportunity to start eliminating Moran's gang, from the bottom up.

With the boss's authorization, McGurn created a six-man team, headed by Fred Burke, with the intention of luring the Gusenbergs, with as many of Moran's other henchmen as possible, into a trap. Burke, a little-known Capone man at the time, invited the brothers to a warehouse meeting, claiming to have many crates of hijacked bootleg whiskey for sale.

Both Capone and McGurn left town to make sure they had watertight alibis. The meeting was to take place on the night of February 14, and, with more of Capone's men placed as strategic lookouts along the surrounding streets, the plan swung into action. Four of McGurn's gang pulled up at the deserted garage, watched by Moran's lookouts, who, deciding the coast was clear, signaled for the seven-strong Gusenberg gang to approach. But after they were inside, two more of McGurn's gang dressed as Chicago police officers approached in a stolen patrol car. Moran's lookouts fled the scene, fearing a police bust, while Capone's remained in place, on standby in case the real police should arrive.

Inside the garage, the fake patrolmen found the suspicious-looking group and ordered them to drop their weapons. All of the gangsters complied, believing their captors were the relatively harmless police force, many of whom were already on the Mob's payroll anyway. However, as they lined up, Capone's four men peeled away, leaving the seven Moran men alone against the wall. Within a split second the gangsters dressed as policemen opened fire using two Thompson submachine guns. They were quickly joined by the remaining gangsters, who pumped bullets into their surprised and defenseless rivals. All seven—James Clark, Adam Heyer, Johnny May, Al Weinshank, Frank and Peter Gusenberg, and Dr. Reinhardt Schwimmer—were left either dead or bleeding to death on the garage floor. The gunfire attracted the attention of residents in the street, but they were soon comforted to see two uniformed policemen in a patrol car "arresting" those responsible. But when neither of the policemen was ever seen again, it led to one of the bloodiest murder mysteries the world has known, and ultimately not a single conviction was ever secured.

The World's Strangest
Unsolved Crimes

∙∙∙

*Three crimes committed over the past century
that continue to baffle police*

One evening in 1974, building workers in Indianapolis employed by the Dowling Construction Company securely locked up the site, leaving a steel demolition ball dangling from a crane more than two hundred feet above the ground. When the operator arrived for work the following morning, he climbed the crane and took his seat in the cab before he noticed the steel ball was missing. It had completely vanished. A thorough search was made and statewide appeals for information were issued. To this day, police officers are puzzled by the theft. No trace of the demolition ball—at nearly three tons in weight, not easy just to slip into one's pocket—has ever been found.

At 10:30 P.M. on the evening of March 9, 1929, Mrs. Locklan Smith heard the sound of screaming coming from the building next door, a small laundry at 4 East 132nd Street in New York. She immediately called the police, who searched the deserted premises until they came across a small, securely locked room at the back. Unable to break in, officers finally managed to gain access by lifting a small boy through a tiny window; he then released the bolts to the door from the inside. In the room lay the body of the laundry owner, Isidore Fink, who had been shot twice in the chest and once through the left hand. Powder burns indicated the gun had been fired at point-blank range, and yet no gun was found in the room.

Isidore had not committed suicide, he had been murdered, although cash in the safe and in Fink's jacket pocket suggested that robbery was not the motive. At first the police believed the murderer must have made his escape through the window, as Isidore always securely bolted the doors from the inside when he worked alone at night. But not only would the window have been too small or awkward to get through (unless the murderer had been a dwarf or a small child), it also did not explain why the killer hadn't simply unbolted the door and walked out through that instead. Others suggested Fink had been shot through the window, but tests proved the powder burns would only show if the gun had been fired from a distance of a few inches, so unless the murderer had twelve-foot arms, they would have to rule that idea out too. No other clue was ever found, and two years after the death of the unfortunate Mr. Fink, the New York police commissioner, Edward P. Mulrooney, was forced to declare the incident an "unsolvable mystery."

At some time between June 28 and July 6, 1907, a person or persons unknown walked into the strong room of Bedford Tower in Dublin Castle and stole the Irish crown jewels, said to be worth £250,000 at the time. Whoever stole them must have had keys, as no locks were broken and there was no sign of a forced entry. How the thieves could have gotten hold of a set of keys is a mystery in itself, as the sole key holder was Sir Arthur Vicars, the Ulster king of arms, who was out of the country at the time. Staff calculated it would have taken between fifteen and twenty minutes to remove the jewels from their individual cases before the thieves made their escape. During this time, none of the four heavily armed guards on duty noticed anything out of the ordinary, and despite a lengthy investigation by Scotland Yard, no trace of the crown jewels has ever been found.

Acknowledgments

Thanks in the first place to all my friends at Harry's Bar in Cape Town (Café del Mar in Camps Bay) for making me so welcome, including James and Melanie Van Vuuren; Tammy Green; the McKenzie sisters, Michelle, Clifflyn, and Heather (what a pool party that was); the twins Melanie and Candice (another story); Freddy Tshibala (Freddy the barman) for the Swahili translation; and of course the big man, Jerry.

At home in South Africa, thanks to Patrick Jones, my man in Cape Town; and to the housekeeper, Miss Ellic, who looked after me so well on the Cape. And special thanks to Margeaux Dawe, Juanita Slabbert, and Madame Zingara—what a last night we all had.

To Beth Lang and the pool boy Ferg Walker for keeping me company for a while—and, obviously, for looking after the pool—and Jon Riley for sitting around doing absolutely nothing over Christmas in Cape Town. Peter Gordon must get a mention again, and I remember why this time, but I can't tell you as it would only embarrass him. Respect to Troy Kyle, a wonderful South African writer who must surely be published soon. Thanks also to Erika Hearle down Guildford way for making my last effort her book of the week.

To Sandra Howgate for the great illustrations, and to the team of Jill Schwartzman and Lea Beresford (editorial), Evan Camfield, Emily Votruba, and Lynn Anderson (copyediting), Richard

Elman (production), Dana Maxson (publicity), and all the bookstores. Where would we be without you.

Finally, and most importantly, to you the readers who enable me to write for a living. To show my appreciation, this book is dedicated to each of you personally, so write your name neatly on the dedication page.

Index

Adams, Polly, 218
airplanes. *See* Bermuda
    Triangle; Cooper, D. B.;
    Miller, Glenn
aliens
    abductions by, 10–11
    Dover Demon as, 89
    Roswell, New Mexico, UFO,
      3–4
    See also beasts and monsters;
      unidentified objects
Allan, Frances, 202
Allan, John, 202–3
Allison, Rupert, 174
Allsop, Jane, 218
Almeida, Avelino de, 105–6
Amado de Melo, João Maria,
    105
Ames, Samuel Brewster, 42
Andrews, Roy Chapman, 124
Angels of Mons, 112
Asher, Jane, 129
Atkinson, Fred W., Jr., 165–66
Aurora, Texas, UFO, 4–11

Baird, William, 211
Baker, Gladys Pearl Monroe,
    186
Bartlett, Bill, 88
Baxter, John, 88
Beard, Joseph, 156

beasts and monsters
    Bigfoot, 25–38
    Dover Demon, 88–89
    Loch Ness Monster, 121–27
    Spring-heeled Jack, 217–20
Beauharnais, Stéphanie de, 119
Beck, Fred, 25, 26
Ben-Menashe, Ari, 173–74, 175
Benoliel, Joshua, 106
Berlitz, Charles, 15
Bermuda Triangle, 12–24
    Flight 19 mystery, 12–13, 14,
      15, 19–21
    geography, 15–17
    magnetic anomalies, 17,
      18–19
    map, 18
    ship mysteries, 14–15, 21,
      22–23
Bigfoot, 25–38
Binding, Tim, 83
Blair, Tony, 113
Blaker, Hugh, 181, 182
Bond, James, 83, 87
Bord, Janet, 32, 33
Bower, Doug, 68, 74
Brabham, Abigail, 89
Briggs, Captain Benjamin
    Spooner, 139, 140, 141,
      142, 143, 144–45, 147,
      149, 150, 151

Burgess, Guy, 86
Burke, Fred, 231
Burns, J. W., 26

Camp, Michael, 198
Campbell, Alex, 123
Campbell, Iain, 99–100
Campbell, William Shears, 129,
    132
Campos, Pedro a, 39
Capet, Hugh, 114
Capet, Louis. *See* Louis-Charles
    of Bourbon, Duke of
    Normandy
Capone, Al "Scarface," 228,
    229–30, 231
Carter, Dr. John F., 205, 210
Catholic Church
    approves Our Lady of Fatima
        visions, 108–9, 110
    in early twentieth-century
        Portugal, 105
cereology, 74
Chalou, George, 162
Charles Louis Edmond de
    Bourbon, 116
Chase, Dorcas, 41, 42, 43,
    44, 46
Chase, Mary Anna Maria,
    40–41, 42, 43
Chase, Thomas, 40–41, 42,
    45, 46
Chase Vault, 39–47
cheating death, 153–57
Chorley, Dave, 68, 74
Christie, Agatha Miller, 48–54
Christie, Archibald, 48, 49, 50,
    51, 52
Circlemakers, 74–75
Clark, Dr. Neil, 125–26
Clark, James, 230, 232
Clark, Thomasina, 43

Clarke, Robert Boucher, 43
Clemm, Maria, 202, 203, 205,
    206–7, 209, 210, 211
Clemm, Virginia, 203, 204
Clemmons, Jack, 192
Columba, Saint, 121, 127
Columbus, Christopher, 18, 24
Combermere, Viscount and
    Lady, 43–44, 45
Conan Doyle, Sir Arthur
    and Chase Vault, 46
    and Cottingley Fairies, 101–2
    and *Mary Celeste,* 142, 151
Conover, David, 187
Cooper, D. B., 55–65
cooping, 211–12
Copperfield, David, 112–13
Costello, Frank, 188
Cottingley Fairies, 101–2
Cotton, Rowland, 43
Coutts, Lady, 177
Coutts, Sir Walter, 171
Cowan, Sir John, 217–18
Cox, Robert, 13
Coy, Janice, 35
Crabb, Lionel "Buster," 80–87
Crabbe, Buster, 80
creatures, mysterious
    Bigfoot, 25–38
    Cottingley Fairies, 101–2
    Dover Demon, 88–89
    Loch Ness Monster, 121–27
    Spring-heeled Jack, 217–20
crop circles, 66–76
crown jewels, Irish, 234–35
Cyclops (ship), 14, 21

da Vinci, Leonardo, as portrait
    painter, 178–83
d'Aragona, Cardinal Luigi,
    179–80
Daumer, Friedrich, 119

d'Avalos, Costanza, 180,
    182–83
De Beatis, Antonio, 179–80
deaths and disappearances,
    mysterious
    Agatha Christie, 48–54
    D. B. Cooper, 55–65
    Lionel "Buster" Crabb, 80–87
    John Dillinger, 77–79
    Eilean Mor lighthouse
        keepers, 90–100
    Louis-Charles of Bourbon,
        Duke of Normandy,
        114–16
    Robert Maxwell, 168–77
    Mary Celeste crew, 138–52
    Glenn Miller, 158–67
    Marilyn Monroe, 184–96
    Edgar Allan Poe, 201–13
    St. Valentine's Day Massacre,
        227–32
deep-sea observation, 221
Deveau, Oliver, 138, 140–41,
    146, 147
Devil's Sea, Japan, 17, 18
Devil's Triangle, 15
    See also *Bermuda Triangle*
Dewis, Joshua, 143
Dietrich, Marlene, 162
Dillinger, John, 77–79
DiMaggio, Joe, 184, 188, 190
DiMaggio, Joe, Jr., 185, 191
disappearances. *See* deaths and
    disappearances, mysterious
DNA testing, of Bigfoot claims,
    35–36
Doherty, Pete, 160
Doug and Dave effect. *See*
    Bower, Doug; Chorley,
    Dave
Dougherty, James, 186
Dover Demon, 88–89

Dowling Construction
    Company, 233
Dublin Castle, Ireland, 234–35
Ducat, James, 91, 92, 93–94, 95

earthquakes. *See* seaquakes
Eastman, Linda, 130, 134
Eden, Anthony, 81, 82, 84, 86
Edwards, Frank, 15
Edwards, John, 160
Eilean Mor lighthouse, 90–100
Einstein, Albert, 184
Elliot, James, 40, 45
Elliott, Nicholas, 84, 86
Engelberg, Dr. (Marilyn
    Monroe's doctor), 192,
    194
Epstein, Brian, 129, 135
Evans, Mary, 4

fairies. *See* Cottingley Fairies
Fatima, Portugal, 103
    See also Our Lady of Fatima
FBI (Federal Bureau of
    Investigation)
    and D. B. Cooper, 57, 58–59,
        60–62, 63, 64–65
    and John Dillinger,
        77–78, 79
    and Marilyn Monroe, 184,
        188, 189
Fersen, Axel de, 116
Finch, Major J., 43
Fink, Isadore, 234
fish, raining, 214–16
Flannan, Bishop, 92
Flannan Islands, 90
"Flannan Isle" (song), 99
Fleming, Ian, 83
Flight 19 (U.S. Navy Avenger
    torpedo bomber), 12–13,
    14, 15, 19–21

Flood, Frederick Solly,
    140–41
flying saucers. *See* spaceships
Fogg (ship). See *V. A. Fogg*
    (ship)
Fosdyk, Abel, 142–43
Francis I, King of France, 180,
    182
frogs, raining, 214–16

Gaddis, Vincent, 15
Galloway, George, 174
Garrett, Dr. Almeida, 107
German autobahn, 136–37
German U-boats, 14
Giancana, Sam, 190
Gibb, Russ, 128, 129
Gibson, William Wilson, 99
Gilling, Andrew, 145
Gimlin, Bob, 29, 31
Giocondo, Francesco di Zanobi
    del, 179, 181, 182
Giocondo, Madonna Lisa del,
    179–83
Goddard, Thomasina, 40, 41,
    42, 43, 46
Gorbachev, Mikhail, 173
Gordon, Stan, 34
Gray, Hugh, 124
Green, John, 26, 28, 29, 31
Greenson, Dr. Ralph, 184, 185,
    188, 189, 191–92, 193,
    194, 196
Griffiths, Frances, 101–2
Gulf Stream, 16–17
Gusenberg, Frank "Tight Lips,"
    228, 231, 232
Gusenberg, Peter, 231, 232

Haines, Joe, 169
Hamilton, Polly, 77
Harcourt, Sir William, 154

Harrison, George, 129
Harvie, Captain, 91, 92
Hauser, Kaspar, 117–20
Haydon, S. E., 4–5, 10
Haynes, Don, 159
Head, Edward William, 145
Heath, Ted, 170
Heironimus, Bob, 30–31
Hermania (ship), 151
Hewes, Hayden, 5–6, 10
Heyer, Adam, 232
hijacking, airplane. *See*
    Cooper, D. B.
Hillary, Edmund, 26
Hiltel, Andreas, 118
Himmelsbach, Ralph, 61–62
Hoch, Jan Ludvik, 168
    See also Maxwell, Robert
Hoover, J. Edgar, 77, 78, 79,
    188, 189
hurricanes, in Bermuda
    Triangle, 19, 21
Hussein, Saddam, 172

illusions, large-scale, 112–13
Inglis, John, 85
Ingram, J. H., 213
insurance industry, and
    Bermuda Triangle, 23, 24
International UFO Bureau,
    5–6
Irish crown jewels, 234–35
Isleworth Mona Lisa, 181, 183

Jack, Donald, 97
Jack, Spring-heeled, 217–20
Jacobs, Arthur, 185–86
John Paul II, Pope, 110–11
Johnson, James, 64
Johnson, John, 138
Johnson, Paul, 34, 35
Jones, E.V.W., 15

Kaiko (submarine), 221, 226
Kairei (submarine), 226
Karl, Grand Duke of Baden, 119
Kaye, Lee, 143
Keating, Lawrence, 143
Kennedy, Joe, 188
Kennedy, John F., 184, 188, 189, 190
Kennedy, Robert F., 184, 186, 188–89, 190, 195
Kerans, J. S., 82
Keyse, Emma Ann, 153
Khrushchev, Nikita, 81, 84
Knowles, Sydney, 81, 82, 83–84
Korablov, Lev Lvovich, 83
Krantz, Dr. Grover, 30, 36, 37
Kryuchkov, Vladimir, 173
Kursk (submarine), 225–26

La Gioconda (painting), 181, 182–83
Lane, John, 49
Lawford, Patricia Kennedy, 188
Lawford, Peter, 188, 190, 191
Lawrence, Jimmy, 79
Lee, John, 153–54
Lembus, Heraclides, 215–16
Lennon, John, 129, 134–35
Leopold I of Baden, 119
Lewis, Isle of, 90, 93
Lewis, John, 214
lighthouse. *See* Eilean Mor lighthouse
Lincoln, Abraham, 113
Linford, Howard, 142, 143
Lipton, Marcus, 82
little green men. *See* aliens
Lloyd's of London, 23, 24
Loch Ness Monster, 121–27
Lomazzo, Giovanni Paolo, 180–81

Long, Greg, 30, 31
Loud, John, 205
Louis XVI, King of France, 114, 116
Louis XVIII, King of France, 115
Louis-Charles of Bourbon, Duke of Normandy, 114–16
Lourdes, 157
Lucas, Nathan, 43, 45
Lucia, Sister, 103, 108, 109, 110–12
Lundberg, John, 74
Lyon, Ben, 187

MacDonald, Mr. (boatswain), 92
Mackay, Mr. and Mrs., 123
MacKenzie, Roderick, 90
Maclean, Donald, 86
Mafia
and Kennedy family, 188–89
and Marilyn Monroe, 184, 188, 190, 196
See also St. Valentine's Day Massacre
Magee, Irvin, 155–56
magnetic forces
and Bermuda Triangle, 17–19
and German autobahn, 136–37
true north versus magnetic north, 17–18
Mallowan, Max, 52
Malone, Tom, 31
Marathon (ship), 151
Marie Antoinette, Queen of France, 114, 116
Marine Sulphur Queen (ship), 14–15, 21

Marshall, Thomas, 91, 92, 93, 94.

Martians. *See* spaceships

Marto, Francisco, 103

Marto, Jacinta, 103

Mary Celeste (ship), 138–52
  conspiracy theories, 140–43
  most credible explanation, 147–51
  troubled history, 144, 151–52

mass delusion, 112–13

Maxwell, Ghislaine, 176

Maxwell, Ian, 177

Maxwell, Kevin, 177

Maxwell, Robert, 168–77
  conspiracy theories, 176–77
  questions about death, 175, 176–77
  ties to Israel, 172, 173–75

May, Johnny, 232

McArthur, Donald, 91, 92, 94, 95, 98–99

McCartney, Paul, 128–35

McCoy, Richard, Jr., 64–65

McGurn, Jack "Machine Gun," 231

McLellan, Robert, 144

Medici, Giuliano de', 180

Miller, Agatha. *See* Christie, Agatha Miller

Miller, Arthur, 184, 188

Miller, Glenn, 158–67

Miller, Helen, 160, 162–63

Miller, Madge, 48

Mills, Bertram, 122–23, 124, 125

Milne, John, 97

Miner, John W., 186, 194

Miracle of the Sun, 105–8

mobsters. *See* Mafia; St. Valentine's Day Massacre

Mocumin, Lugne, 121

Mona Lisa. *See* Giocondo, Madonna Lisa del

Mona Lisa (painting), 178–83

Monroe, Marilyn, 184–96
  autopsy report, 186, 193
  death, 185–86, 190–96
  relationship with Kennedy brothers, 184, 186, 188–89, 190, 195

Mons, Belgium, Battle of, 112

monsters. *See* beasts and monsters

Moore, Joseph, 91–92, 93, 97

Moran, Dr. (attends Poe), 207, 209–10, 211

Moran, George "Bugs," 227, 228, 229, 231

Morehouse, David Reed, 138, 140, 145

Morgan, John R., 159

Morris, Philip, 30

Morrison, N. H., 213

Mortenson, Norma Jeane. *See* Monroe, Marilyn

Mount St. Helens, 63

Muirhead, Robert, 92, 93–97

Mulrooney, Edward P., 234

Murdoch, Rupert, 170

Murray, Eunice, 185, 191, 192, 194

Napier, John, 30, 31–32

Napoleon, 115, 119

natural/supernatural mysteries
  Bermuda Triangle, 12–24
  crop circles, 66–76
  German autobahn, 136–37
  Our Lady of Fatima, 103–13
  raining frogs and fish, 214–16
  See also seaquakes

Naundorff, Charles-Guillaume, 115–16
Neele, Nancy, 49, 50, 52
Newcomb, Pat, 185
Niven, David, 158, 159, 162, 167
Noguchi, Dr. Thomas, 186
Noll, Rick, 32–33
Norgay, Tenzing, 26
Northwest Orient Airlines, 56, 57
Nyrop, Donald, 57

Oats, Rollie, 6
Orderson, Thomas, 43
organized crime. *See* Mafia
Ostman, Albert, 28–29
Our Lady of Fatima, 103–13
    initial appearances, 103, 104
    and Miracle of the Sun, 105–8
    possible explanations, 112–13

paranormal events. *See* natural/supernatural mysteries
Parker, Gilman, 151–52
Parker, John Nutting, 144
Patterson, Roger, 29, 30, 31
Pemberton, John, 143
Pergamon Press, 170–71
Philby, Kim, 86
Piano Man, 197–200
Pinto Coelho, Dr. Domingos, 106–7
pirates
    as explanation for some Bermuda Triangle mysteries, 22
    ruled out in *Mary Celeste* case, 141

Pius XII, Pope, 110
Platt, John, 224
Poe, David, Sr. (grandfather), 213
Poe, Edgar Allan, 201–13
    "cooping" theory, 211–12
    death, 201–2, 209
    mysteries of last trip, 202, 205–7
Poe, Neilson (cousin), 210–11, 212–13
Portugal
    earthquake destroys Lisbon, 148
    Our Lady of Fatima, 103–8, 110
Powell, Henry, 39
Powell, John, 39
Proctor, Judge, 5, 6
Profumo, John, 184
Prohibition era, 227
Puerto Rico Trench, 17
Pulitzer, Dr. Henry F., 182
Purvis, Melvin, 77
Purvis, Will, 154–56

quantum physics, 34, 35

raining frogs and fish, 214–16
Raphael, sketches Leonardo's painting of Mona Lisa, 179
Richardson, Albert, 145
Ross, William, 94
Ruah, Judah, 106
Ryan's Tavern, Baltimore, 201, 205

Sage, Anna, 77, 79
Sand, George X., 15
Santilli, Ray, 3–4
Santos, Artur de Oliveira, 103, 104, 105

Santos, Lucia Rosa dos, 103–4
  See also Lucia, Sister
Sargasso Sea, 12, 16
  See also Bermuda Triangle
sasquatch, defined, 26
  See also Bigfoot
Savoy, Grand Duke of, 181
Scales, Lucy, 218
Schaffner, Florence, 56–57
Schwimmer, Dr. Reinhardt, 232
Scott, William, 57–58
seaquakes
  and disappearance of Eilean
    Mor lighthouse keepers,
    100
  and disappearance of *Mary
    Celeste* crew, 149–50, 151
  in vicinity of Azores, 148–49
Shag Harbor, Nova Scotia,
    222–23
Shamir, Yitzhak, 172
Shaw, Fred, 165
Shelton, Elmira, 204, 205, 206
ships
  Cyclops, 14, 21
  Hermania, 151
  Marathon, 151
  Marine Sulphur Queen,
    14–15, 21
  Mary Celeste, 138–52
  V. A. Fogg, 22–23
Shipton, Eric, 25–26
Sinatra, Frank, 184, 190
Sinclair, Sir John, 82
Smith, Mrs. Locklan, 233
Snelling, Harold, 101
Snodgrass, Dr. Joseph, 201,
    208–9
spaceships
  Aurora, Texas, UFO, 4–11
  and Bermuda Triangle,
    14, 15

Roswell, New Mexico alien,
    3–4
Spring-heeled Jack, 217–20
St. Valentine's Day Massacre,
    227–32
Starr, Ringo, 129
Steinberg, Saul, 170–71
Stevens, Charlie, 4
Stevens, Mary, 218
Stevenson, David, 93
Stonehenge, 69–70
submarines, 221, 223,
    224–25
Suez Canal, 81, 84
supernatural mysteries. *See*
    natural/supernatural
    mysteries

Taintor, Will, 89
Taylor, Charles, 12–13, 19–20
Taylor, Zachary, 212
Thurman, Derek, 166
Townshend, Peter, 200
Traynor, Jack, 156–57
true north versus magnetic
    north, 17–18

U-boats, 14
UFOs. *See* spaceships;
    unidentified objects
underwater seaquakes, 100,
    149–50, 151
underwater vessels, 221–26
unidentified objects
  Aurora, Texas, UFO, 4–11
  underwater versions,
    221–26
unidentified submersible
    objects (USOs), 221–26
unsolved crimes and mysteries
  Chase Vault disturbances,
    39–47

disappearance of Eilean Mor
  lighthouse keepers, 90–100
disappearance of *Mary Celeste*
  crew, 138–52
laundry owner's murder, 234
missing demolition ball,
  233–34
theft of Irish crown jewels,
  234–35
USOs (unidentified
  submersible objects),
  221–26

V. A. Fogg (ship), 22–23
Valentine's Day. *See* St.
  Valentine's Day Massacre
Vanunu, Mordechai, 174
vapor flashes, 149–50
Vasari, Giorgio, 179, 182
Vicars, Sir Arthur, 234
Virgin Mary, 103, 104, 109–11
  See also Our Lady of Fatima
von Hochberg, Countess, 119
von Tucher, Baron, 119

Wabc, John, 73
Walden, Brian, 176

Waldron, Elizabeth, 40
Waldron, Thomas, 40
Walker, Joseph, 201, 209
Wallace, Ray, 26–27, 28
Weeks, Dr. Patrick, 78
Wehrs, Carl, 136–37
Weinshank, Al, 232
Wellington, Duke of, 115, 219
Wells, H. G., 75
Wetherell, Marmaduke, 123
Wheyleigh, John, 73
Whiteside, Dr. Lowell, 148
Wickens, Laurie, 222
Wild, David, 224
Wilson, Harold, 170
Wilson, Robert Kenneth, 125
Winchester, James H., 144, 147,
  151
Worley, G. W., 14
Wright, Elsie, 101–2
Wright, John, 138
Wright, Wilbur, 161–62

Yeltsin, Boris, 173

Zarski, Tom, 128, 135
Zwerkin, Joseph, 83

ALBERT JACK has become something of a publishing phenomenon with his huge bestsellers *Red Herrings and White Elephants* and *Shaggy Dogs and Black Sheep,* clocking up hundreds of thousands of sales. Fascinated by discovering the truth behind the world's great stories, Albert has become an expert at explaining the unexplained. And besides that, he loves a good story . . .

When not engaged in research, he lives somewhere between Guildford, U.K., and Cape Town, South Africa, where he divides his time between fast living and slow horses, neat vodka and untidy pubs.